LIFE MOVEMENTS IN PLANTS

LIFE MOVEMENTS IN PLANTS

VOL. 1

J. C. BOSE

ANMOL PUBLICATIONS
New Delhi (India)

ANMOL PUBLICATIONS
4378/4B, Ansari Road
Daryaganj, New Delhi-110 002

Reprint 1989
Reprinted 1993

ISBN 81-7041-172-6 (Set)
ISBN 81-7041-173-4 (Vol. 1)

PRINTED IN INDIA
Published by J.L. Kumar for Anmol Publications, New Delhi-110 002 and
Printed at Mehra Offset Press, Delhi-110 006.

BOSE INSTITUTE

I DEDICATED the Institute to the Nation on the 30th November 1917. The fuller investigation of the many ever-opening problems of the nascent science, which includes both Life and Non-Life, is among the main purposes of the Institute. In adding a large auditorium, I have sought permanently to associate the advancement of knowledge with the widest possible civic and public diffusion of it; and this without any academic limitations, henceforth to all races and languages, to both men and women alike, and for all time coming.

It is my further wish that, as far as the limited accommodation would permit, the facilities of this Institute should be available to workers from all countries. In this I am attempting to carry out the traditions of my country, which, so far back as twenty-five centuries ago, welcomed all scholars from different parts of the world within the precincts of its ancient seats of learning, at Nalanda and at Taxilla.

The present time may be regarded as inopportune for any new undertaking, when a great tragedy is hanging over human destiny. But it is at such a crisis that men learn to discriminate the Real from the Unreal, so that they can dedicate themselves to the pursuit of Truth which is eternal.

Since the opening of the Institute I have received much encouragement and help from different provinces of India in this national undertaking. My grateful thanks are due to the gentlemen whose first donations have formed the nucleus of an Endowment Fund, and also to Mr. P. C. Lyon, the retired Minister of Education, Government of Bengal, through whose representation the Government have been pleased to make generous grants in furtherance of research. I am indebted to His Excellency Lord Ronaldshay, Governor of Bengal, for the grant of land for necessary expansion and for his interest in the permanence of the Institute. The Government of India and the Government of Bengal have also sanctioned six Scholarships for Research students of the Institute.

The Government has also kindly undertaken the publication of the Transactions of the Institute to be issued twice a year. Much time had to be given last year in the organization of the Institute. In spite of this a long series of investigations on the phenomenon of plant movements have been completed in the course of the year. Of these the first series is given in the present volume.

J. C. BOSE.

October 1918.

BOSE INSTITUTE.

Director.

Sir Jagadis Chunder Bose, Kt., M.A., D.Sc., C.S.I., C.I.E.

Scholars and Research Assistants.

Guruprasanna Das, L.M.S.
Surendra Chandra Das, M.A.
Narendra Nath Neogi, M.Sc.
Basiswar Sen, B.Sc.
Jyotiprakash Sircar, B.Sc., M.B., Ch.B.
Narendra Nath Sen Gupta.
Satyendra Nath De, M.Sc.
Lalit Mohan Mukhopadhya, B.Sc. (*Nawroji Scholar*).
Satyendra Chandra Guha, M.Sc.
Abani Nath Mitter, *Superintendent*.

Honorary Secretary.

Sudhansu Mohan Bose, M.A., LL.B.

Trustees of the Institute.

Board of Managers.

The Director of the Institute ... } *Ex-officio*
The Superintendent of the Institute } Members
Hon'ble Sir Nil Ratan Sircar.
Satis Ranjan Das.
Sudhansu Mohan Bose.
Dr. Banwari Lal Chaudhuri.
Lady Bose.
Dr. Prankrishna Acharya.
A representative of Research Scholars.

Visitors.

His Excellency Lord Ronaldshay, G.C.I.E.
Maharaja Sir Manindra Chandra Nandy, K.C.I.E.

Donors.

His Highness the Maharaja Gaekwar of Baroda.
S. R. Bomanji, Bombay.
Mulraj Khatau, Bombay.
Maharaja Sir Manindra Chandra Nandy, K.C.I.E.

Honorary Members.

Lord Rayleigh, O.M., D.Sc., D.C.L., LL.D., F.R.S.
Professor S. H. Vines, F.R.S., University of Oxford.
Professor P. C. Ray, D.Sc., C.I.E.
P. C. Lyon, C.S.I., Fellow, Oriel College, Oxford.
President Stanley Hall, Clark University.
Professor R. A. Millikan, University of Chicago.
Professor R. A. Harper, Columbia University.
Professor Patrick Geddes, University of St. Andrews.

THE VOICE OF LIFE.*

I DEDICATE to-day this Institute—not merely a Laboratory but a Temple. The power of physical methods applies for the establishment of that truth which can be realized directly through our senses, or through the vast expansion of the perceptive range by means of artificially created organs. We still gather the tremulous message when the note of the audible reaches the unheard. When human sight fails, we continue to explore the region of the invisible. The little that we can see is as nothing compared to the vastness of that which we cannot. Out of the very imperfection of his senses man has built himself a raft of thought by which he makes daring adventures on the great seas of the Unknown. But there

* Inaugural address in dedication of the Bose Institute, November 30, 1917.

are other truths which will remain beyond even the super sensitive methods known to science. For these we require faith, tested not in a few years but by an entire life. And a temple is erected as a fit memorial for the establishment of that truth for which faith was needed. The personal, yet general, truth and faith whose establishment this Institute commemorates is this; that when one dedicates himself wholly for a great object, the closed doors shall open, and the seemingly impossible will become possible for him.

Thirty-two years ago I chose teaching of science as my vocation. It was held, that by its very peculiar constitution, the Indian mind would always turn away from the study of Nature to metaphysical speculations. Even had the capacity for inquiry and accurate observation been assumed present, there were no opportunities for their employment; there were no well-equipped laboratories nor skilled mechanicians. This was all too true. It is for man not to quarrel with circumstances but bravely accept them; and we belong to that race and dynasty who had accomplished great things with simple means.

FAILURE AND SUCCESS.

This day twenty-three years ago, I resolved that as far as the whole-hearted devotion and faith of one man counted, that would not be wanting, and within six months it came about that some of the difficult problems connected with electric waves found their solution in my Laboratory, and received high appreciation from the leading physicists. The Royal Society honoured me by publishing my discoveries and offering, of their own accord an appropriation from the special Parliamentary grant for the advancement of knowledge. That day the closed gate suddenly opened, and I hoped that the torch that was then lighted would continue to burn brighter and brighter. But man's faith and hope require repeated testing. For five years after this the progress was uninterrupted; yet when the most generous and wide appreciation of my work had reached almost the highest point there came a sudden and unexpected change.

LIVING AND NON-LIVING.

In the pursuit of my investigations I was unconsciously led into the border region of physics and physiology, and was amazed to find boundary lines vanishing and points of contact emerge between the realms of the Living and Non-Living. Inorganic matter was found anything but inert; it also was a-thrill under the action of multitudinous forces that played on it. A common reaction seemed to bring together metal, plant and animal under a general law. They all exhibited the phenomena of fatigue and depression, together with possibilities of recovery and of exaltation, yet also that of permanent irresponsiveness which is associated with death. I was filled with awe at this great generalization; and it was with great hope that I announced my results before the leading scientific society. The results were so unexpected that they provoked incredulity. There were also misgivings about the inherent bent of the Indian mind towards mysticism and unchecked imagination. But in India this burning imagination which can extort new order out of a mass of apparently contradictory facts, is also held in check by the habit of meditation. It is this restraint which confers the power to hold the mind in pursuit of truth, in infinite patience, to wait and reconsider, to experimentally test and repeatedly verify.

It is but natural that there should be prejudice even in science, against all innovations; but there were added other difficulties which it was impossible to remove from this isolating distance. Thus no conditions could have been more hopeless than those which confronted me for the next twelve years. It is necessary to make this brief reference to this period of my life; for one who devote himself to the search of truth must realize that for him there awaits no easy life, but one of unending struggle. It is for him to cast his life as an offering, regarding gain and loss, success and failure, as one. Yet in my case this long persisting gloom was suddenly lifted. My scientific deputation in 1914, from the Government of India, gave the opportunity of giving demonstrations of my results before the leading scientific societies of the world. This led to a very generous appreciation of my work and the recognition of the importance of Indian contribution to the advancement of the world's science. My own experience told me how heavy, sometimes even crushing, are the difficulties which confront an inquirer here in India; yet it made me stronger in my determination, that I shall make the path of those who are to follow me less arduous, and that India is never to relinquish what has been won for her after years of struggle.

THE TWO IDEALS.

What is it that India is to win and maintain? Has her own history and the teaching of the past prepared her for some temporary and quite subordinate gain? There are at this moment two complementary and not antagonistic ideals before the country. There is first, the individualistic ideal of winning success in all affairs, of securing material efficiency and of satisfaction of personal ambition. These are necessary, but by themselves cannot ensure the life of a nation. Such material activities have brought in the West their fruit in accession of power and wealth. There has been a feverish rush even in the realm of science for exploiting applications of knowledge, not so often for saving as for causing destruction. In the absence of some power of restraint, civilization is now trembling in an unstable poise on the brink of ruin. Some complementary ideal there must be to save man from that mad rush which must end in disaster. He has followed the lure and excitement of some insatiable ambition, never pausing for a moment to think of the ultimate object for which success was to serve as a temporary incentive. He forgot that far more potent than competition, was mutual help and co-operation in the scheme of life. And in his country through millenniums, there always have been some who, beyond the immediate and absorbing prize of the hour, sought for the realization of the highest ideal of life—not through passive renunciation, but through active struggle. The weakling who has refused the conflict, having acquired nothing, has nothing to renounce. He alone who has striven and won can enrich the world by giving away the fruits of his victorious experience. In India such examples of constant realization of ideals through work have resulted in the formation of a continuous living tradition. And by her latent power of rejuvenescence she has readjusted herself through infinite transformations. Thus while the soul of Babylon and the Nile valley have transmigrated, ours still remains vital and with capacity of absorbing what time has brought, and making it one with itself.

The ideal of giving, of enriching, in fine, of self-renunciation in response to the highest call of humanity is the other and complementary ideal. The motive power for this is not to be found in personal ambition but in the effacement of all littlenesses, and uprooting of that ignorance which regards anything as gain which is to be purchased at others' loss. This I know, that no vision of truth can come except in the absence of all sources of distraction, and when the mind has reached the point of rest. And for my disciples, I call on those very few, who will devote their whole life with strengthened character and determined purpose to take part in that infinite struggle to win knowledge for its own sake and see truth face to face.

ADVANCEMENT AND DIFFUSION OF KNOWLEDGE.

The work already carried out in my Laboratory on the response of matter, and the unexpected revelations in plant life, foreshadowing the wonders of the highest animal life, have opened out very extended regions of inquiry in Physics, in Physiology, in Medicine, in Agriculture and even in Psychology. These inquiries are obviously more extensive than those customary either among physicists or physiologists, since demanding interests and aptitudes hitherto more or less divided between them. In the study of Nature, there is a necessity of the dual viewpoint, this alternating yet rhythmically unified interaction of biological thought with physical studies, and physical thought with biological studies. The future worker with his freshened grasp of physics, his fuller conception of the inorganic world, as indeed thrilling with "the promise and potency of life", will redouble his former energies of work and thought. Thus he will be in a position to winnow the old knowledge with finer sieves, to re-search it with new enthusiasm and subtler instruments. And thus with thought and toil and time he may hope to bring fresher views into the old problems. His handling of these will be at once more vital and more kinetic, more comprehensive and unified.

The further and fuller investigation of the many and ever-opening problems of the nascent science which includes both Life and Non-Life are among the main purposes of the Institute I am opening to-day; in these fields I am already fortunate in having a devoted band of disciples, whom I have been training for the last ten years.

The advance of science is the principal object of this Institute and also the diffusion of knowledge. We are here in the largest of all the many chambers of this House of Knowledge—its Lecture Room. In adding this large auditorium, I have sought permanently to associate the advancement of knowledge with the widest possible civic and public diffusion of it; and this without any academic limitations, henceforth to all races and languages, to both men and women alike, and for all time coming.

The lectures given here will announce, to an audience of some fifteen hundred people, the new discoveries made here, which will be demonstrated for the first time before the public. We shall thus maintain continuously the highest aim of a great seat of learning by taking an active part in the advancement and diffusion of knowledge. Through the regular publication of the Transactions of the Institute, these Indian contributions will reach the whole world. The discoveries made will thus become public property. No patents will ever be taken. The spirit of our national culture demands that we should for ever be free from the desecration of utilizing knowledge for personal gain.

It is my further wish, that as far as the limited accommodation would permit, the facilities of this Institute should be available to workers from all countries. In this I am attempting to carry out the traditions of my country, which so far back as twenty-five centuries ago, welcomed all scholars, from different parts of the world, within the precincts of its ancient seats of learning, at Nalanda and at Taxilla.

THE SURGE OF LIFE.

With this widened outlook, we shall not only maintain the highest traditions of the past but also serve the world in nobler ways. We shall be at one with it in feeling the common surgings of life, the common love for the good, the true, and the beautiful. In this Institute, this Study and Garden of Life, the claim of Art has not been forgotten, for the artist has been working with us, from foundation to pinnacle, and from floor to ceiling of this very Hall. And beyond that arch, the Laboratory merges impercep tibly into the Garden, which is the true laboratory for the study of Life. There the creepers, the plants and the trees are played upon by their natural environments,—sunlight and wind, and the chill at midnight under the vault of starry space. There are other surroundings also, where they will be subjected to chromatic action of different lights, to invisible rays, to electrically charged atmosphere. Everywhere they will transcribe in their own script the history of their experience. From his lofty point of observation, sheltered by the trees, the student will watch this panorama of life. Isolated from all distractions, he will learn to attune himself with Nature; the obscuring veil will be lifted and he will gradually come to see how community throughout the great ocean of life outweighs apparent dissimilarity. Out of discord he will realize the great harmony.

THE OUTLOOK.

These are the dreams that wove a network round my wakeful life for many years past. The outlook is endless, for the goal is at infinity. The realization cannot be through one life or one fortune, but through the co-operation of many lives and many fortunes. The possibility of a fuller expansion will depend on very large endowments. But a beginning must be made, and this is the genesis of the foundation of this Institute. I came with nothing and shall return as I came; if something is accomplished in the interval, that would indeed be a privilege. What I have I will offer, and one who had shared with me the struggles and hardships that had to be faced, has wished to bequeath all that is hers for the same object. In all my struggling efforts I have not been altogether solitary; while the world doubted, there had been a few, now in the City of Silence, who never wavered in their trust.

INDIA'S SPECIAL APTITUDES IN CONTRIBUTION TO SCIENCE.

The excessive specialization in modern science has led to the danger of losing sight of the fundamental fact that there can be but one truth, one science, which includes all the branches of knowledge. How chaotic appear the happenings in Nature! Is Nature a Cosmos, in which the human mind is some day to realize the uniform march of sequence, order and law? India through her habit of mind is peculiarly fitted to realize the idea of unity, and to see in the phenomenal world an orderly universe. This trend of thought led me unconsciously to the dividing frontiers of different sciences and shaped the course of my work in its constant alternations between the theoretical and the practical, from the investigation of the inorganic world to that of organized life and its multifarious activities of growth, of movement, and even of sensation. On looking over a hundred different lines of investigations carried on during the last twenty-three years, I now discover in them a natural sequence. The study of electric waves led to the devising of methods for the production of exceedingly short electric waves which served as a bridge over the gulf between visible and invisible lights; from this followed accurate investigation on the optical properties of invisible waves, the determination of the refractive powers of various opaque substances, the discovery of effect of air film on total reflection, and the polarizing properties of strained rocks and of electric tourmalines. The invention of a new type of self-recovering electric receiver made of galena was the fore-runner of crystal detectors. In physical chemistry the detection of molecular change in matter under electric stimulation led to a new theory of photographic action. The theory of stereo-chemistry found strong support by the production of two kinds of artificial molecules, which, like the two kinds of sugar, rotated the polarized electric wave either to the right or to the left. Again the 'fatigue' of my receivers led to the discovery of sensitiveness inherent in matter as shown by its electric response. It was next possible to study this response in its modification under changing environment, of which its exaltation under stimulants and its abolition under poisons are among the most astonishing outward manifestations. And as a single example of the many applications of this fruitful discovery, the characteristics of an artificial retina gave a clue to the unexpected discovery of "binocular alternation of vision" in man;—each eye thus supplements its fellow by turns, instead of acting as a continuously yoked pair, as hitherto believed.

PLANT LIFE AND ANIMAL LIFE.

In natural sequence to the investigation of the response in "inorganic" matter, has followed a prolonged study of the activities of plant life as compared with the corresponding functioning of animal life. But since plants for the most part seem motionless and passive, and are indeed limited in their range of movement, special apparatus of extreme delicacy had to be invented, which should magnify the tremor of excitation and also measure the perception period of a plant to a thousandth part of a second. Ultra-microscopic movements were measured and recorded, the length measured being often smaller than a fraction of a single wave-length of light. The secret of plant life was thus for the first time revealed by the autographs of the plant itself. This evidence of the plant's own script removed the prevailing error which divided the vegetable world into sensitive and insensitive. The remarkable performance of the 'Praying' Palm Tree of Faridpur, which bows, as if to prostrate itself, every evening, is only one of the latest instances which show that the supposed insensibility of plants and still more of rigid trees is to be ascribed to wrong theory and defective observation. Investigations which I have carried out show that all plants, even the trees, are fully alive to changes of environment; they respond visibly to all stimuli, even to the slight fluctuations of light caused by a drifting cloud. This series of investigations has completely established the fundamental identity of life reactions in plant and animal, as seen in a similar periodic insensibility in both, corresponding to what we call sleep; as seen in the death-spasm, which takes place in the plant as in the animal. This unity in organic life is also exhibited in that spontaneous pulsation which in the animal is heart-beat; it appears in the identical effects of stimulants, anæsthetics and of poisons in vegetable and animal tissues. This physiological identity in the effect of drugs is regarded by leading physicians as of great significance in the scientific advance of medicine; since here we have a means of testing the effect of drugs under conditions far simpler than those presented by the patient, far subtler too, as well as more humane than those of experiments on animals.

Growth of plants and its variations under different treatment is instantly recorded by my Crescograph. Authorities expect that this method of investigation will advance practical agriculture; since for the first time we are able to analyse and study separately the conditions which modify the rate of growth. Experiments, which would have taken months and their results vitiated by unknown changes, can now be carried out in a few minutes.

Returning to pure science, no phenomena in plant life are so extremely varied or have yet been more incapable of generalization than the "tropic" movements, such as the twining of tendrils, the heliotropic movements of some towards and of others away from light, and the opposite geotropic movements of the root and shoot, in the direction of gravitation or away from it. My latest investigations have led to the establishment of a single fundamental reaction which underlies all these effects so extremely diverse.

Finally, I may say a word of that other new and unexpected chapter which is opening out from my demonstration of nervous impulse in plants. The speed with which the nervous impulse courses through the plant has been determined; its nervous excitability and the variation of that excitability have likewise been measured. The nervous impulse in plant and in man is found exalted or inhibited under identical conditions.

A question long perplexing physiologists and psychologists alike is that concerned with the great mystery that underlies memory. But now through certain experiments I have carried out, it is possible to trace "memory impressions" backwards even in inorganic matter, such latent impressions being capable of subsequent revival. Again the tone of our sensation is determined by the intensity of nervous excitation that reaches the central perceiving organ. It would be possible to change the tone or quality of our sensation, if means could be discovered by which the nervous impulse would become modified during transit. Investigation on nervous impulse in plants has led to the discovery of a controlling method, which was found equally effective in regard to the nervous impulse in animals.

Thus the lines of physics, of physiology, and of psychology converge and meet. And here will assemble those who would seek oneness amidst the manifold. Here it is that the genius of India should find its true blossoming.

The thrill in matter, the throb of life, the pulse of growth, the impulse coursing through the nerve and the resulting sensations, how diverse are these and yet how unified! How strange it is that the tremor of excitation in nervous matter should not merely be transmitted but transmuted and reflected like the image on a mirror, from a different plane of life, in sensation and in affection, in thought and in emotion! Of these which is more real, the material body or the image which is independent of it? Which of these is undecaying, and which of these is beyond the reach of death?

It was a woman in the Vedic times, who, when asked to take her choice of the wealth that would be hers for the asking, inquired whether that would win for her deathlessness. Many a nation had risen in the past and won the empire of the world. A few buried fragments are all that remain as memorials of the great dynasties that wielded the temporal power. There is, however, another element which finds its incarnation in matter, yet transcends its transmutation and apparent destruction: that is the burning flame born of thought which has been handed down through fleeting generations.

Not in matter, but in thought, not in possessions or even in attainments, but in ideals, are to be found the seed of immortality. Not through material acquisition but in generous diffusion of ideas and ideals can the true empire of humanity be established. Thus to Asoka to whom belonged this vast empire, bounded by the inviolate seas, after he had tried to ransom the world by giving away to the utmost, there came a time when he had nothing more to give, except one-half of an *Amlaki* fruit. This was his last possession, and his anguished cry was that since he had nothing more to give, let the half of the *Amlaki* be accepted as his final gift.

Asoka's emblem of the *Amlaki* will be seen on the cornices of the Institute, and towering above all is the symbol of the thunderbolt. It was the Rishi Dadhichi, the pure and blameless, who offered his life that the divine weapon, the thunderbolt, might be fashioned out of his bones to smite evil and exalt righteousness. It is but half of the *Amlaki* that we can offer now. But the past shall be re-born in a yet nobler future. We stand here to-day and resume work to-morrow, so that, by the efforts of our lives and our unshaken faith in the future we may all help to build the greater India yet to be.

CONTENTS

PART I.

RESPONSE OF PLANT ORGANS.

I.—THE PROBLEM OF MOVEMENT IN PLANTS.

II.—THE "PRAYING" PALM TREE.

III.—ACTION OF STIMULUS ON VEGETABLE TISSUES.

IV.—THE DIURNAL VARIATION OF EXCITABILITY IN *MIMOSA*.

V.—RESPONSE OF PETIOLE-PULVINUS PREPARATION OF *MIMOSA*.

VI.—CONDUCTION OF EXCITATION IN PLANTS.

VII.—ELECTRIC CONTROL OF EXCITATORY IMPULSE.

VIII.—EFFECT OF INDIRECT STIMULUS ON PULVINATED ORGANS.

IX.—MODIFYING INFLUENCE OF TONIC CONDITION ON RESPONSE.

PART II.

GROWTH AND ITS RESPONSIVE VARIATIONS.

X.—THE HIGH MAGNIFICATION CRESCOGRAPH FOR RESEARCHES ON GROWTH.

XI.—EFFECT OF TEMPERATURE ON GROWTH.

XXII.—EFFECT OF CHEMICAL AGENTS ON GROWTH.

XIII.—EFFECT OF VARIATION OF TURGOR AND OF TENSION ON GROWTH.

XIV.—EFFECT OF ELECTRICAL STIMULUS ON GROWTH.

PAGE.

XV.—EFFECT OF MECHANICAL STIMULUS ON GROWTH.

XVI.—ACTION OF LIGHT ON GROWING ORGANS.

XVII.—EFFECT OF INDIRECT STIMULUS ON GROWTH.

XVIII.—RESPONSE OF GROWING ORGANS IN STATE OF SUB-TONICITY.

XIX.—RESUMPTION OF AUTONOMOUS PULSATION AND OF GROWTH UNDER STIMULUS.

XX.—ACTION OF LIGHT AND WARMTH ON AUTONOMOUS ACTIVITY.

XXI.—A COMPARISON OF RESPONSES IN GROWING AND NON-GROWING ORGANS.

ILLUSTRATIONS.

I.—THE PROBLEM OF MOVEMENT IN PLANTS

By

PROF. SIR J. C. BOSE.

THE phenomenon of movement in plants under the action of external stimuli presents innumerable difficulties and complications. The responding organs are very different: they may be the pulvini of the 'sensitive' or those of the less excitable leguminous plants; the petioles of leaves, which often act as pulvinoids; and organs of plants in a state of active growth.

Taking first the case of the pulvinus of *Mimosa*, we find that it responds to mechanical stimulation, to constant electric current, to induction shock, to the action of chemical agents, to light, and to warmth as differentiated from thermal radiation. The reactions induced by these agents may be similar or dissimilar. An identical agent, again, may give rise to movements which are not merely different, but sometimes even of diametrically opposite characters. Certain organs, for example, direct themselves towards light, others away from it. Some plants close their leaflets on the approach of darkness, in the so-called position of 'sleep'; apparently similar 'sleep' movement is induced in others by the action of the midday sun.

In *Mimosa*, the responsive movement is brought about by a sudden diminution of turgor in the pulvinus. But very little is definitely known about the responsive reaction in growing organs. Thus in a tendril, one-sided contraction causes a shortening of the concave side and a sudden increase of growth on the convex. No explanation of this difference has hitherto been forthcoming. Under the action of light of different intensities a growing organ may approach the source of light, or place itself at right angles or move away from it. Again under the identical stimulus of gravity, the root moves downwards, and the shoot upwards. The sign of response in different organs thus changes, apparently without any reason. It is thus seen, that there is hardly any responsive movement that has been observed of which an example directly to the contrary may not be found. For this reason it has appeared hopeless to unify these very diverse phenomena, and there has been a tendency towards a belief that it was not any definite physiological reaction, but the individuality of the plant that determines the choice of its movement.

The complexities which baffle us may, however, arise from the combination of factors whose individual reactions are unknown to us. I shall show, for example, how the movement of a pulvinus under a given stimulus is determined by the point of application, direct stimulus producing one effect, and indirect the diametrically opposite. The normal reaction is again modified by the tonic condition of the plant. There is again the likelihood of the presence of other modifying factors. It is clear how very different the results would become by the permutation and combination of these diverse factors.

For a comprehensive study of the phenomenon of plant movement, it is therefore necessary to investigate in detail the effect of a given stimulus under definite changes of the environmental condition. With regard to a given stimulus we have to determine the effects of intensity, of duration, and of the point of application. The investigation has to include the effects exhibited not merely by the pulvinated but also by growing organs. As a result of such a comprehensive study, it may perhaps be possible to discover some fundamental reaction operative in bringing about the responsive movement in all plant organs.

I shall, in the course of the following series of Papers, describe the different apparatus by which the movement of pulvinated organ and its time-relations are automatically recorded. In a growing organ the induced movement under stimulus is brought about by the change in its rate of growth. That the change is solely due to the particular stimulus can only be assured by strict maintenance of constancy of external conditions, during the period of experiment; this constancy can, in practice, be secured only for a short time. The necessity for shortening the period of experiment also arises from a different consideration; for numerous and varied are the stimulating and mechanical interactions between neighbouring organs. These effects, however, come into play after a certain lapse of time. They may be eliminated by reduction of the period of experiment.

In order to shorten the period of experiment for the study of growth movements, the rate of growth has to be very highly magnified, so as to determine the absolute rate and its variations in the course of a minute or so. I shall in a subsequent Paper give full account of an apparatus I have been able to devise, by which it is possible to record automatically the rate of growth magnified many thousand times.

I stated that anomalies of plant movements would disappear, if we succeeded in carrying out in detail investigations of effects of the different individual factors in operation. In illustration of this I shall, in the first Paper of the series, give an account of the mysterious movement of the 'Praying' Palm of Faridpur, and describe the investigations by which the problem found its solution.

II.—THE "PRAYING" PALM TREE

By

SIR J. BOSE,

Assisted by

NARENDRA NATH NEOGI, M.Sc.

PERHAPS no phenomenon is so remarkable and shrouded with greater mystery as the performances of a particular Date Palm near Faridpur in Bengal. In the evening, while the temple bells ring calling upon people to prayer, this tree bows down as if to prostrate itself. It erects its head again in the morning, and this process is repeated every day of the year. This extraordinary phenomenon has been regarded as miraculous, and pilgrims have been attracted in large numbers. It is alleged that offerings made to the tree have been the means of effecting marvellous cures. It is not necessary to pronounce any opinion on the subject; these cures may be taken as effective as other faith-cures now prevalent in the West.

This particular Date Palm, *Phœnix dactylifera*, is a full-grown rigid tree, its trunk being 5 metres in length and 25 cm. in diameter. It must have been displaced by storm from the vertical and is now at an inclination of about 60° to the vertical. In consequence of the diurnal movement, the trunk throughout its entire length is erected in the morning, and depressed in the afternoon. The highest point of the trunk thus moves up and down through one metre; the 'neck,' above the trunk, is concave to the sky in the morning; in the afternoon the curvature disappears, or is even slightly reversed. The large leaves which point high up against the sky in the morning are thus swung round in the afternoon through a vertical

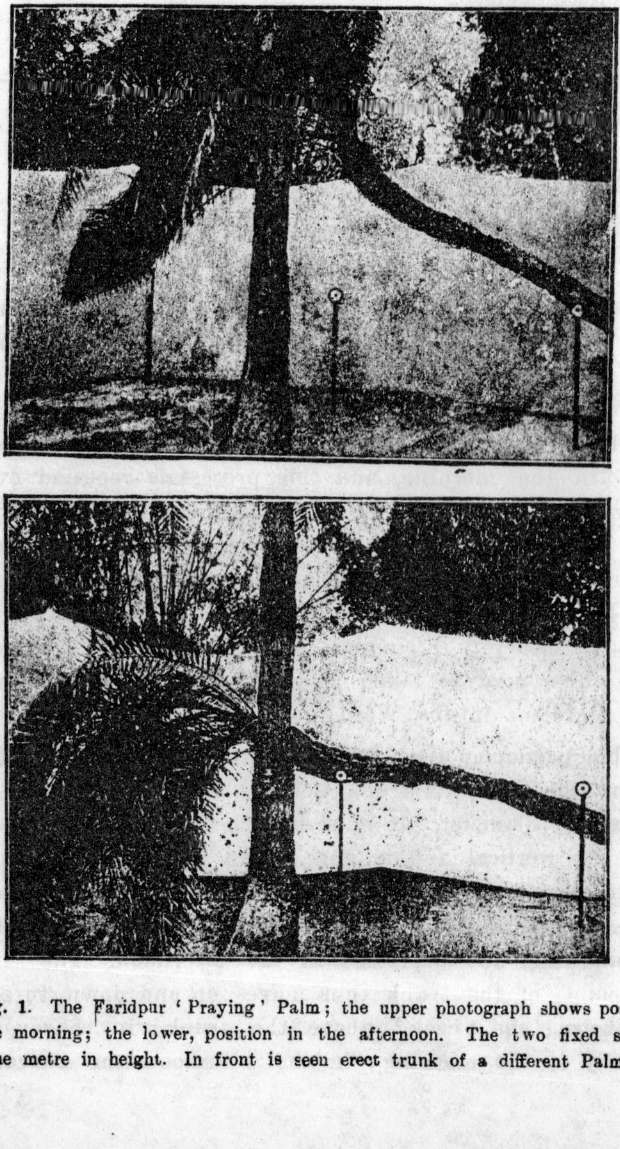

Fig. 1. The Faridpur 'Praying' Palm; the upper photograph shows position in the morning; the lower, position in the afternoon. The two fixed stakes are one metre in height. In front is seen erect trunk of a different Palm.

distance of about five metres. To the popular imagination the tree appears like a living giant, more than twice the height of a human being, which leans forward in the evening from its towering height and bends its neck till the crown of leaves press against the ground in an apparent attitude of devotion (Fig. 1). Two vertical stakes, each one metre high, give a general idea of the size of the tree and movements of the different parts of the trunk.

For an investigation in elucidation of this phenomenon it was necessary:—

1. To obtain an accurate record of the movement of the tree day and night, and determine the time of its maximum erection and fall.
2. To find whether this particular instance of movement was unique, or whether the phenomenon was universal.
3. To discover the cause of the periodic movement of the tree.
4. To find the reason of the remarkable similarity between the diurnal movement of the tree, and the diurnal variation of moto-excitability in *Mimosa pudica*.
5. To determine the relative effects of light and temperature on the movement.
6. To demonstrate the physiological character of the movement of the tree.
7. To discover the physiological factor whose variation determines the directive movement.

THE RECORDING APPARATUS.

I shall now describe the principle and construction of my recording apparatus (Fig. 2) seen attached to a horizontally growing stem of *Mimosa pudica*. When used to trace the movement of the palm tree, a reducing device is employed to keep the record within the plate. A lever, R[1], records the movement of the attached tree or plant on a moving plate of smoked glass. The plate is not in contact with the

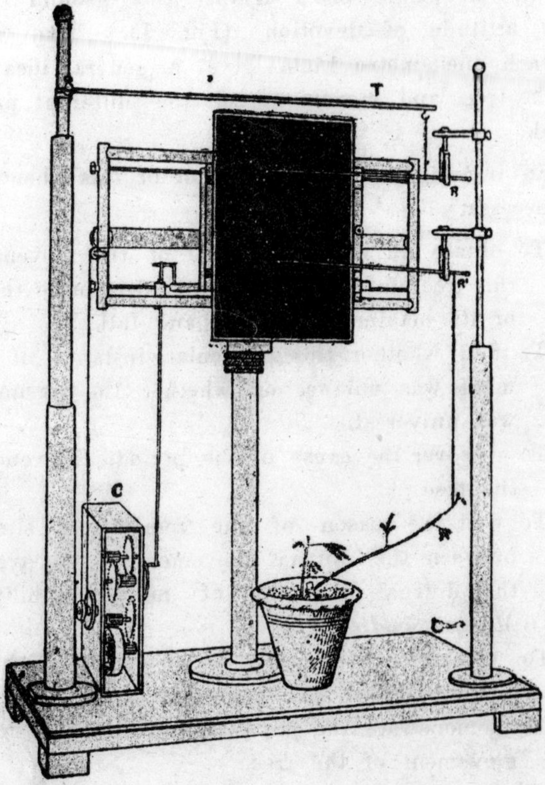

Fig. 2. Apparatus for automatic record of movement of trees and plants; T, differential metallic thermometer; R, recording lever for temperature; R[1], for recording plant movement; C, clock-work for oscillation of recording plate. The same clock-work moves plate laterally in 24 hours.

tip of the recording lever, but separated from it by a distance of about 3 mm. A special oscillating device, actuated by clock-work, C, makes the plate move forwards and backwards. The forward movement brings about a momentary contact of the recording tip with the smoked plate inscribing a dot. These single dots are made at intervals of 15 minutes; at the expiration of the hour, however, contact is made three times in rapid succession, printing a thick dot. It is thus easy to determine the movement of the tree at all times of the day and night. A second lever, R, placed above, gives on the same plate, thermographic record of the diurnal variation of temperature. For this I use a differential thermometer, T, made of a compound strip of brass and steel. Curvature is induced by the differential expansion of the two pieces of metal. The up or down movement of the free end of the compound strip is further magnified by the recording lever. This arrangement was extremely sensitive and gave accurate record of variation of temperature. By the forward movement of the oscillating plate two dots are made at the same time,—one for the temperature and the other for the corresponding movement of the tree. As the two recorders do not move vertically up or down, but describe a circle, the dots vertically one above the other may not correspond as regards time. Any possibility of error in calculation is obviated by the fact that the thick dots in both the records are made every hour, and the subsequent thin dots at intervals of 15 minutes.

A difficulty arose at the beginning in obtaining sanction of the proprietor to attach the recorder to the tree. He was apprehensive that its miraculous power might disappear by profane contact with foreign-looking instruments. His misgivings were removed on the assurance that the instrument was made in my laboratory in India, and that it would be attached to the tree by one of my assistants, who was the son of a priest.

From results of observation it is found that the tree moves through its entire length; the fall of the highest point of the trunk is one metre. The movement is not

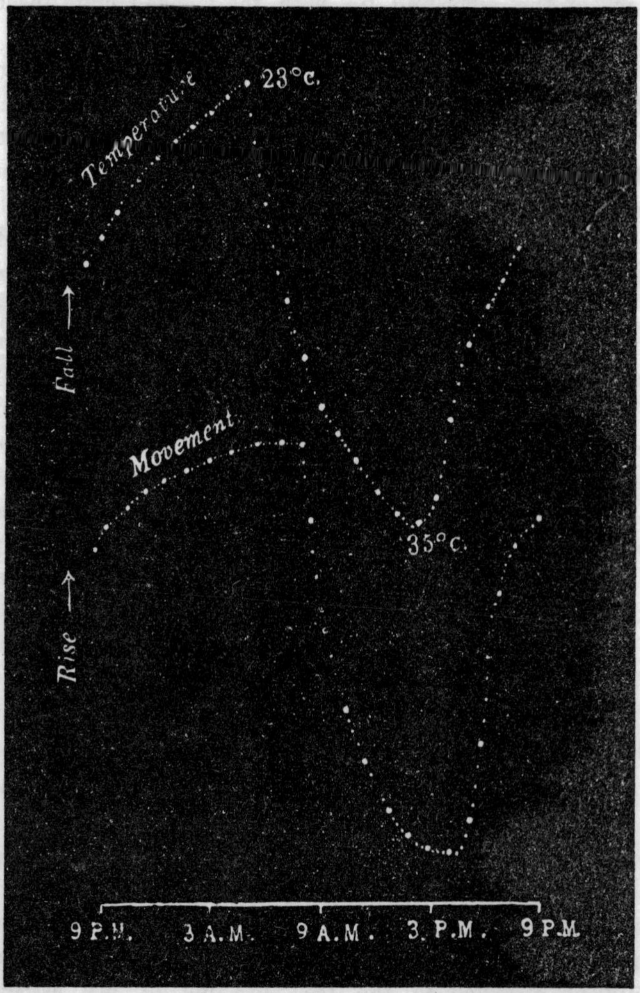

Fig. 3. Record of diurnal movement of the 'Praying' Palm (*Phœnix dactylifera*). Thermographic curve for 24 hours commencing at 9 in the evening is given in the upper record; the corresponding diurnal curve of movement of the tree is given in the lower. Successive dots at intervals of 15 minutes; thick dots at intervals of an hour.

passive, but an active force is exerted; the force necessary to counteract this movement is equivalent to the weight of 47 kilograms: in other words, the force is sufficient to lift a man off the ground. But far greater force would be required to restrain the change of curvature of the neck of the hard and rigid tree.

Before entering into the investigation of the cause of periodic movement I shall give a general account of its characteristics. A casual observation would lead one to conclude that the tree lifted itself at sunrise and prostrated at sunset. But continuous record obtained with my recorder attached to the upper part of the trunk shows that the tree was never at rest, but in a state of continuous movement, which underwent periodic reversals (Fig. 3). The tree attained its maximum erection at 7 in the morning, after which there is a rapid movement of fall. The down movement reached its maximum at 3-15 P.M., after which it was reversed and the tree erected itself to its greatest height at 7 next morning. This diurnal periodicity was maintained day after day.

UNIVERSALITY OF TREE MOVEMENT.

The next question which I wished to investigate was whether the movement of the particular Faridpur tree was a unique phenomenon. It appeared more likely that similar movement would, under careful observation, be detected in all trees. The particular palm tree was growing at a considerable inclination to the vertical; the movement of the tree and its leaves became easily noticeable, since the ground afforded a fixed and striking object of reference. In a tree growing more or less erect, the movement, if any, would escape notice, since such movements would be executed with only the empty space as the background.

Experiment 1.—Believing the phenomenon to be universal I experimented with a different Date Palm that was growing at my research station at Sijbaria on the

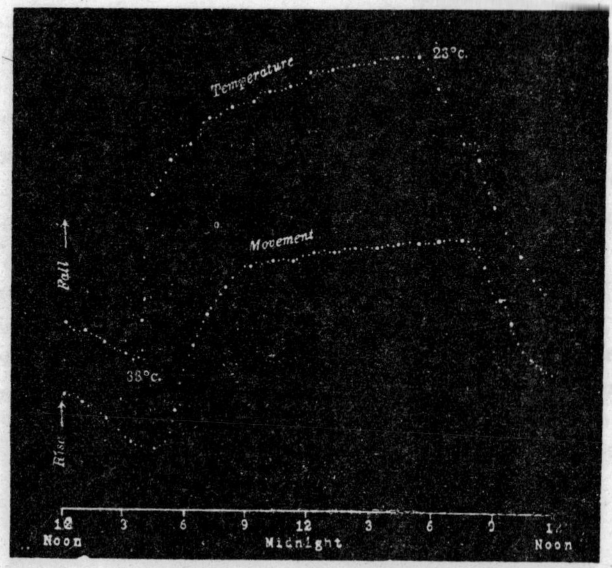

Fig. 4. Record of the Sijbaria Palm from noon for 24 hours. Successive dots, at intervals of 15 minutes.

Ganges, situated at a distance of about 200 miles from Faridpur. The surrounding conditions were very different. The tree was much younger; it was 2 metres in height and inclined 20° to the vertical. The curve obtained with this tree (Fig. 4) was very similar to that of the Faridpur Palm, though the extent to the movement was much reduced. The tree attained the highest erect position at 7-15 A.M. and the lowest at 3-45 P.M. Hence the movement of the Faridpur Palm is not a solitary phenomenon.

THE CAUSE OF PERIODIC MOVEMENT.

The recurrent daily movement of the tree must be due to some diurnal changes in the environment,—either the recurrent changes of light and darkness, or the diurnal changes of temperature. These changes synchronise to a certain extent; for, as the sun rises, light appears and the temperature begins to rise. It is therefore difficult to discriminate the effect of light from that of temperature. The only satisfactory method of discrimination would have been in the erection of a large structure with screens to cut off light. The effect of fluctuation of temperature under constant darkness would have demonstrated the effect of one agent without complication arising from the other. Unfortunately screening the tree was impracticable. I shall presently describe other experiments where the action of light was completely excluded.

The curve of movement of the tree, however, affords us material for correct inference as regards the relative effects of light and temperature. The experiment was commenced in March; light appeared at about 5 A.M., the sunrise being at 6-15 A.M.; the sun set at 6-15 P.M., and it became dark by 7 P.M. The incident light would be the most intense at about noon; after this it would decline continuously till night time. If the movement was due to light, its climax, either in up or down movement, would be reached at or about noon, and the opposite climax at midnight. But instead of this we find (Fig. 3) the up-movement reaching its highest point not at noon, but at 7 in the morning; after this the fall is rapid and continuous, and the lowest position was reached not in the evening but at 3-15 P.M. The fluctuation of light has, therefore, little to do with the movement of the tree.

Turning next to the element of variation of temperature we are at once struck by the fact that the curve of movement of the tree is practically a replica of the thermographic curve (Fig. 3). The *fall* of temperature is seen to induce a *rise* in the tree and *vice versâ*. There is a lag in the turning points of the two curves; thus while temperature began to rise at 6 A.M., the tree did not begin to fall till 7 A.M. There is in this case a lag of an hour; but the *latent* period may, sometimes, be as long as three hours. The delay is due to two reasons; it must take some time for the thick trunk of the tree to attain the temperature of the surrounding, and secondly, the physiological inertia will delay the reaction. As a result of other investigations, I find that the induced effect always lags behind the inducing cause. It is interesting in this connection to draw attention to the parallel phenomenon, which is described below, of lag in the variation of sensibility of *Mimosa* in response to variation of temperature. In this case the lag was found to be about three hours. Returning to the Palm, the tree continues to fall in the forenoon with rising temperature. At about 2-30 P.M. the temperature was at its maximum after which it began to decline; the movement of the tree was not reversed into erection till after 3-15 P.M., the lag being now 45 minutes nearly.

I may state here that the movement of the tree is not primarily affected by the periodicity of day and night, but by variation of temperature. In spring and in early summer the rise of temperature during the early part of the day and the fall of the temperature from afternoon to next morning, are regular and continuous; the corresponding movements of the tree are also regular. But at other seasons, owing to the sudden change of direction of the wind, the fluctuations of temperature are irregular. Thus at night there may be a sudden rise, and in the earlier part of the day sudden fall of temperature. And the record of movement of the tree is found to follow these fluctuations with astonishing fidelity, the rise of temperature being followed by a fall of the tree and *vice versâ*. That the movement is determined by the temperature variation is exhibited in a striking manner in Fig. 4, where, between 8 and 9 A.M., a common twitch will be noticed in the two curves.

While trying to obtain some clue to the mysterious movement of the tree, my attention was strongly attracted by certain striking similarities which the record of the movement of the tree showed to the curve of the diurnal variation of moto-excitability, of the pulvinus of *Mimosa pudica*, an account of which will be found in a subsequent Paper of the series.*

PERIODIC MOVEMENT OF TREES AND DIURNAL VARIATION OF MOTO-EXCITABILITY IN *MIMOSA PUDICA*.

The excitability of the main pulvinus of *Mimosa pudica* I find does not remain constant during the 24 hours, but undergoes a striking periodic change. At certain hours of the day, the excitability is at its maximum ; at a different period it practically disappears. The period of insensibility is about 7 A.M., which, strangely enough, is also the time when the palm tree attains its maximum height. At about 3 in the afternoon the excitability of *Mimosa* reaches its climax, and this is the time when the head of the palm tree bends down to its lowest position. For the determination of the periodic variation of excitability of *Mimosa* I devised a special apparatus by which an electric stimulus of constant intensity

* *See* also Bose—Diurnal Variation of Moto-Excitability in *Mimosa*—Annals of Botany, Vol. XXVII, No. CVIII, October, 1913.

was automatically applied to the plant every hour of the day and night, the responsive moment being recorded at the same time. The amplitude of responsive fall of leaf under uniform stimulus gave a measure of excitability of

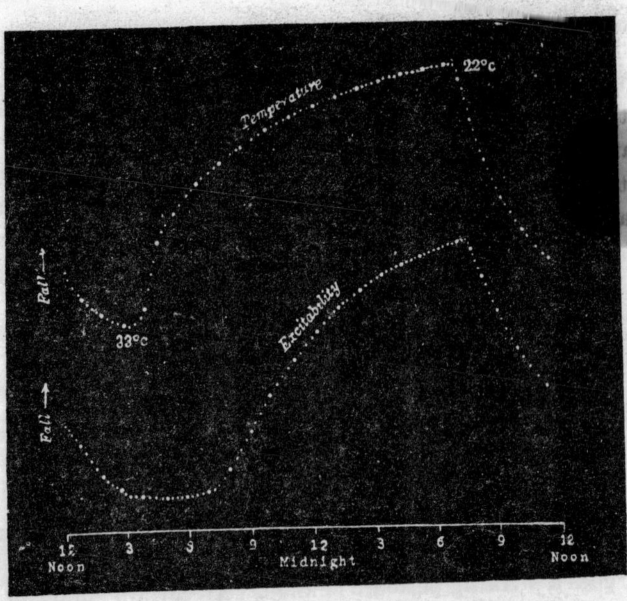

Fig. 5. Curve of variation of moto-excitability of *Mimosa pudica*. The upper curve gives variation of temperature and the lower, the corresponding variation of excitability.

the leaf at any particular moment In the lower curve of Fig. 5 is given the record of diurnal variation of excitability of *Mimosa*. Comparison of this figure with Figs. 3 and 4, will show the remarkable resemblance between the curves of diurnal movement of the Palm tree, and of diurnal variation of moto-excitability of *Mimosa*. The excitability of *Mimosa* reached its maximum at about 3 in the afternoon, when the Palm was at its lowest position. After this hour excitability fell continuously till 7 or 8 next morning. Corresponding to this is the continuous erection of the Palm from its lowest position at 3 P.M. to the highest between 7 and 8 A.M. Still more remarkable is the modifying influence of variation of temperature on the diurnal curve of excitability in *Mimosa*, and the diurnal curve of movement of the Palm. This will be quite evident from the inspection of the temperature curves in Figs. 4 and 5.

I have shown elsewhere* that the variation of moto-excitability of the pulvinus of *Mimosa* is a physiological function of temperature. The remarkable similarity between the diurnal variation of moto-excitability of *Mimosa* and diurnal movement of the Palm is due to the fact that both are determined by the physiological action of temperature. I shall presently describe experiments, which will establish the physiological character of the movement of the tree in response to changes of temperature.

The records that have been given show that it is the diurnal variation of temperature, and not of light that is effective in inducing the periodic movement of the tree. Further experiments will be given in support of this conclusion.

RELATIVE EFFECTS OF LIGHT AND TEMPERATURE.

As regards the possibility of light exerting any marked influence on the movement of the Palm tree, I have shown from study of time-relations of the movement, that this could not be the case. Moreover, it is impossible for light to reach the living tissue through the thick layer of bark

* Bose—"Irritability of Plants," d. 60.

that surrounds the tree. That the effect of light is negligible will appear from the accounts of following experiments, where the possibility of the effect of changing intensity of light is excluded by maintaining the plant in constant darkness, or in constant light.

The employment of the large Palm was obviously impracticable in these investigations. I, therefore, searched for other plant-organs in which the movement under variation of temperature was similar to that of the Date Palm. I found that the horizontally spread leaves of vigorous specimens of *Arenga saccharifera* growing in a flower pot executed movements which were practically the same as that of the Faridpur tree. The leaf moved downwards with rise of temperature and *vice versâ*.

There are many practical advantages in working with a small specimen. It can easily be placed under glass cover or taken to a glass house, thus completely eliminating the troublesome disturbance caused by the wind.

Diurnal movement in continued darkness: Experiment 2.— The plant was placed in a dark room and records taken continuously for three days. These did not differ in any way from the normal records taken in a glass house under daily variation of light and darkness. Exposure of plant to darkness for the very prolonged period of a week or more, undoubtedly interferes with the healthy photo-tonic condition of the plant. But such unhealthy condition did not make its appearance in the first few days.

PHYSIOLOGICAL CHARACTER OF THE MOVEMENT.

There may be a misgiving that the movement of the tree might be due to physical effect of temperature. If the upper strip of a differential thermometer be made of the more expansible brass and the lower of iron, the compound strip bends down with the rise of temperature. Similarly the movement of the tree might be due to the upper half being physically more expansible. It would have been possible to discriminate the physical from the physiological action by causing the death of the tree; in that case physical movement would have persisted, while the physiological action would have disappeared. As this test was not practicable, I tried the effect of physiological depression on the periodic movement of the leaf of *Arenga saccharifera*.

Effect of Drought: Experiment 3.—In Fig. 6 is given a series of records of movement of the leaf-stalk of *Arenga*, first under normal condition, afterwards under increasing drought, brought about by withholding water. The uppermost is the thermographic record which remained practically the same for successive days. Below this are records of movement of the leaf (*a*) under normal condition, (*b*) after withholding water for three days, and (*c*) after deprivation for seven days. It will be noticed how the extent of movement is diminished under increasing physiological depression brought on by drought. On the seventh day, the responsive movement disappeared, there being now a mere fall of the leaf, which was slow and continuous. After this I supplied the plant with water and the periodic movement was in consequence nearly restored to its original vigour.

Effect of poison: Experiment 4.—In another experiment the normal diurnal record with the leaf was taken and the plant was afterwards killed by application of poisonous solution of potassium cyanide. The diurnal movement was found permanently abolished at the death of the plant.

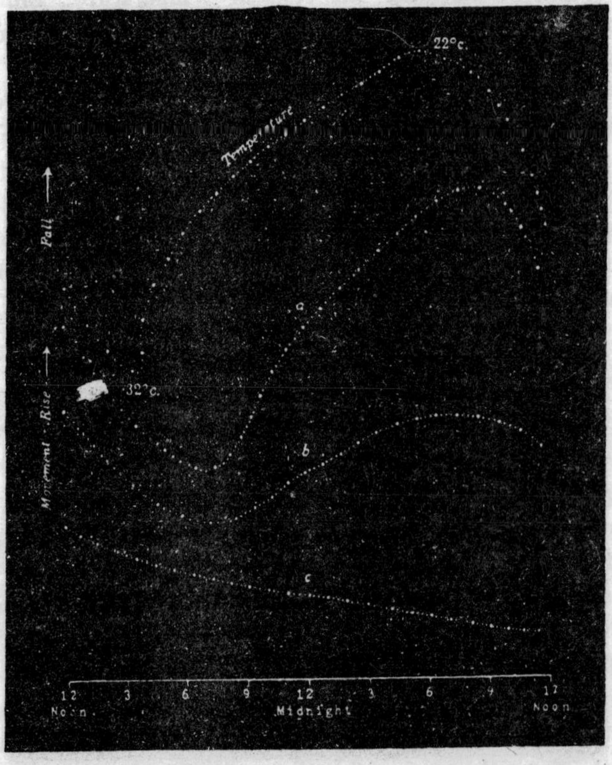

Fig. 6. Effect of physiological depression on diurnal movement of the petiole of *Arenga saccharifera*. The uppermost curve exhibits variation of temperature, (*a*), normal diurnal curve, (*b*), modification after 3 days' and (*c*) after 7 days' withholding of water.

These experiments conclusively prove that the periodic movement of the leaf-stalk induced by variation of temperature is a physiological phenomenon, and from analogy we are justified in drawing the inference that the movement of the Faridpur tree is also physiological. The question, however, was finally settled by the unfortunate death of the tree which occurred the other day, nearly a year after I commenced my investigations. While presiding at my lecture on the subject, His Excellency Lord Ronaldshay, the Governor of Bengal, announced that a telegram had just reached him from his officer at Faridpur that "the palm tree was dead, and that its movements had ceased."

Since my investigation with the Faridpur 'Praying, Palm, I have received information regarding other Palms, which exhibit movements equally striking. One of the trees is growing by the side of a tank, the trunk of the tree being inclined towards it. The up-lifted leaves of this tree are swung round in the afternoon and dipped into the water of the tank.

The movement of the tree has been shown to be brought about by the physiological action of temperature variation; in other words the diurnal movement of the 'Praying' Palm is a THERMONASTIC PHENOMENON. I have found various creeping stems, branches and leaves of many trees, exhibit this particular movement of fall with a rise of temperature, and *vice versâ*. Such movements. I shall, for the sake of convenience, distinguish as belonging to the *negative type*.

Having found that the temperature is the modifying cause, the next point of inquiry relates to the discovery of the force, whose varying effects under changing temperature induces the periodic movement. I shall, in this connection, first discuss the various tentative theories that may be advanced in explanation of the movement.

TRANSPIRATION AND DIURNAL MOVEMENT.

It may be thought that the fall of the tree during rise of temperature may be due to passive yielding of the tree to its weight, there being increased transpiration and general loss of turgor at high temperature. I shall, however, show that the diurnal movement persists in the absence of transpiration.

Diurnal movement in absence of transpiration: Experiment 5.—In the leaf of *Arenga saccharifera*, I found that the petiole was the organ of movement. I cut off the transpiring lamina and covered the cut end with collodion flexile. The plant was now placed in a chamber saturated with moisture. The petiole continued to give records of its diurnal movement in every way similar to the record of the intact leaf. In another experiment with the water plant, *Ipoemia reptans*, immersed in water, the normal diurnal movement was given by the plant, where there could be no question of variation of turgor due to transpiration. (See also *Expt.* 7.)

In the diurnal movement of the 'Praying' Palm the concave curvature of the rigid neck in the morning, became flattened or slightly convex in the afternoon. The force necessary to cause this is enormously great, and could on no account result from the passive yielding to the weight of the upper part of the tree.

From the facts given above it will be seen that the diurnal movement is not brought about by variation in transpiration. I now turn to another phenomenon which appeared at first to have some connection with the movement of the tree. Kraus found that the tissue tensions of a shoot exhibit a daily periodicity. He, however, found that between 10°C. and 30°C., variation of temperature had no effect on the daily period. But as regards the diurnal movement of the tree, it is the temperature which is the principal factor. Kraus also found a daily variation of bulk in different plant-organs; this variation of bulk is connected with transpiration, for the removal of the transpiring leaves arrested this variation. But the periodic movement of the tree, as we have seen, is independent of transpiration.

Millardet observed a daily periodicity of tension in *Mimosa pudica*. He found that maximum tension occurs before dawn; the petiole becomes erected, the movement being upwards or towards the tip of the stem. Tension decreases during the day, and reaches a minimum early in the evening; in correspondence with this is the fall of the petiole, the movement being away from the tip of the stem.* If the plant were placed upside down the periodic movement of the petiole in relation to the stem will evidently remain the same, but become reversed in space. Maximum tension in the morning will make the petiole approach the tip of the stem, *i.e.*, the movement will be *downwards* instead of upwards as in the normal position. The experiment described below will show that the diurnal movement induced by variation of temperature is not reversed by placing the plant in an inverted position.

Diurnal movement in inverted position: Experiment 6.—I took a vigorous specimen of *Arenga saccharifera* growing in a pot, and took its normal record, which as explained before exhibited down-movement during rise, and an up-movement during fall of temperature. The plant was now held inverted, the upper side of the petiole now facing the earth. The diurnal curve of movement should now show an inversion, if that movement was solely determined by the anisotropy of the organ. But the record did not exhibit any such inversion. After being placed upside down, the leaf did not, on the first day, show any diurnal movement; there was, on the other hand, a continuous down-movement

* Vines.—'Physiology of Plants,' 1886, pp. 405 and 543.

on account of the fall of the leaf by its own weight. But in the course of 24 hours the leaf readjusted itself to its unaccustomed position, and became somewhat erected under the action of geotropic stimulus. After the attainment of this new state of geotropic equilibrium, the leaf gave a very pronounced record of its diurnal movement which did not show any reversal; the inverted leaf continued to exhibit the same characteristic movements as in the normal position, that is to say, a down movement during rise, and an up-movement during fall of temperature. As the plant in the inverted position did not show any reversal of the periodic curve, it is clear that the diurnal movement is determined by the modifying influence of temperature on the physiological reaction of the plant to some external stimulus which is constant in direction. I shall presently show that it is the constant geotropic stimulus modified by the action of temperature, which determines the diurnal movement of the tree.

This will be better understood if I refer once more to certain characteristics in the movement of the "Praying" Palm. The neck of the tree was seen to be concave in the morning. The physiological effect of raising temperature is virtually to oppose or neutralise the geotropic curvature as seen in the flattening or slight reversal of curvature in the afternoon. Similarly, various plant organs, growing at an inclination to the vertical, are subjected to geotropic action, and thus assume different characteristic angles. This state of equilibrium is not static but may better be described as dynamic; for it will be shown that this state of geotropic balance is upset in a definite way, by variation of temperature.

That geotropism is an important factor in the diurnal movement is supported by the fact that the Sijbaria Palm with an inclination of 20° to the vertical exhibited a daily movement which was only moderate in extent. But the Faridpur Palm growing at an inclination of 60° was subjected more effectively to geotropic action, and exhibited movements which were far more pronounced. I shall now proceed to describe crucial experiments which will demonstrate the effect of change of temperature on geotropic curvature.

EFFECT OF VARIATION OF TEMPERATURE ON GEOTROPIC CURVATURE.

In the instances of diurnal movement already described the trees or their leaves were already at an inclination to the vertical. I now took a radial and erect shoot of *Basella cordifolia* growing in a pot and laid it horizontally for two weeks. The procumbent stem curved up and attained a state of equilibrium under the action of geotropic stimulus.

Diurnal curve of Basella cordifolia : *Experiment 7.*— The plant was completely immersed in a vessel of water, and its diurnal curve recorded. This resembled in all essentials the diurnal curve of the Palm ; the slight deviation was due to the fact that owing to difference in the season (August) the temperature maximum was attained at 12-25 P. M., and the minimum at 6 A. M. The geotropic curvature was reduced to its minimum at the maximum temperature, and *vice versâ.* As in the case of the Palm so also in the procumbent stem of *Basella* there was a physiological lag, which was 50 minutes in the morning and about the same in the afternoon. The free end of the stem thus exhibited a diurnal movement up and down. The temperature, as stated before, began to rise from 6 A. M. and the down-movement commenced 50 minutes later, *i.e.*, at 6-50 A. M. The temperature, after reaching

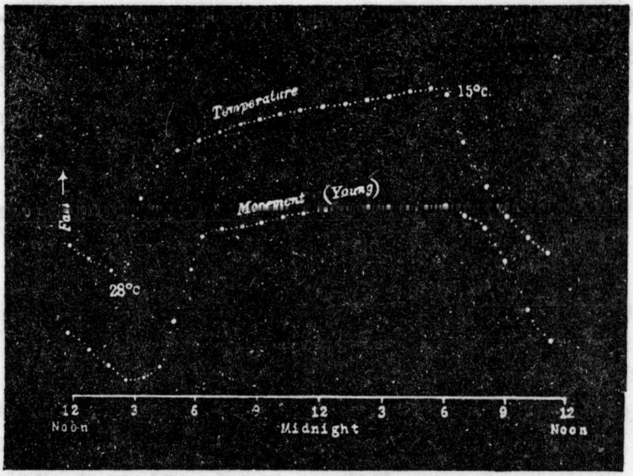

Fig. 7. Diurnal curve of movement of procumbent young stem of *Mimosa pudica*. Successive dots at intervals of 15 minutes.

the maximum, began to fall at 12-25 P. M., and the previous movement of fall of the stem was arrested and reversed into an erectile movement shortly after 1 P. M. There are thus two "turning points," one at 7 A. M., and the other at about 1 P.M.; at these periods the movement of the plant remains more or less arrested for more than half-an-hour.

I obtained records of similar diurnal movements with various procumbent or creeping stems. Figure 7 gives the diurnal record of the procumbent stem of a young specimen of *Mimosa pudica*.

The experiment that has just been described shows clearly that geotropic curvatures of stems is opposed, or neutralised to a greater or less extent, during rise of temperature, and this antagonistic reaction is removed during the fall of temperature. The diurnal movement of the plant completely immersed under water shows once more that transpiration has little to do with the diurnal movement.

REVERSAL OF NATURAL RHYTHM.

The diurnal rhythm of up and down movement in the particular specimen *Basella* had become established under the daily variation of temperature. I now attempted to reverse this rhythm by artificial variation of temperature. The plant was placed in water in a rectangular metallic vessel which was placed within a second outer vessel. The plant could thus be subjected, without any mechanical disturbance, to variation of temperature, by circulating warm or cold water in the outer vessel. In order to reverse the natural rhythm I subjected the plant to the action of falling temperature at the "turning" point at 7 A. M., at a time when the plant would have undergone a down-movement under the daily rise of temperature. Conversely the plant was subjected to the action of rising temperature at the second "turning" point at 1 P. M., when the movement under diurnal fall of temperature would have been one of erection.

Effect of fall of temperature: Experiment 8.—As stated before the experiment was carried out in the morning; ice cold water was circulated in the outer chamber, the fall of temperature was in this case sudden, and there was an almost immediate responsive movement. This appeared anomalous, since the latent period of response to slow variation of temperature was found from the diurnal curve to be as long as 50 minutes.

As a result of further investigations I found that variation of temperature produces two different effects which may be distinguished as transient and persistent. Sudden variation of temperature affects the superficial tissue, and gives rise to a transient reaction, while it takes a long time for temperature variation to react on the geotropically active tissue in the interior. The persistent effect therefore takes place after a latent period from one to three hours according to the thickness of the plant.

The persistent effect of rise of temperature is a movement downwards, that of fall of temperature is a movement upwards. These definite reactions will be seen exhibited in Figs. 8 and 9. The plant was stationary at the turning point in the morning hence the curve at first was horizontal. The temperature was gradually lowered through 5 C., from 29 C, to 24 C. in the course of five minutes and maintained at the lower temperature. There was no immediate effect, but after a latent period of 65 minutes the plant responded by a movement of erection. The natural movement at this period of the day would have been one of fall, but artificial change of temperature in the opposite direction effectively reversed the normal diurnal movement. The latent period for this reverse

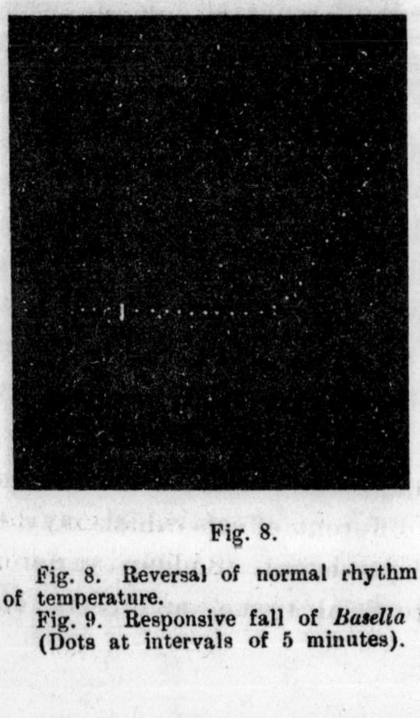

Fig. 8.

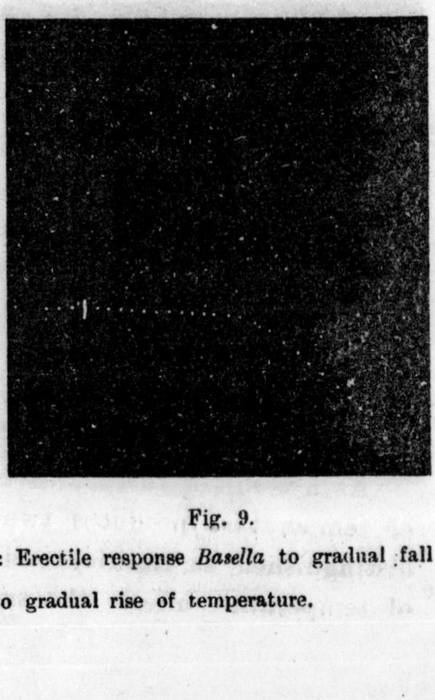

Fig. 9.

Fig. 8. Reversal of normal rhythm : Erectile response *Basella* to gradual fall of temperature.
Fig. 9. Responsive fall of *Basella* to gradual rise of temperature.
(Dots at intervals of 5 minutes).

movement is, as stated before, 65 minutes as against 50 minutes in the normal diurnal movement. The increase in the latent period is probably due to the added physiological inertia in reversing the normal rhythm.

Effect of rise of temperature: Experiment 9.—The temperature was raised through 5°C at the second turning point at 1 P.M. After a latent period of 50 minutes the plant began to rise steadily (Fig. 9) thus exhibiting once more the reversal of its normal diurnal movement.

From the experiments described above it will be seen that the movement of the Palm, and of other organs growing at an inclination to the vertical, is brought about by the action of temperature in modifying the geotropic curvature. The ever present tendency of geotropic movement is opposed or helped by the physiological reaction induced by rise and fall of temperature respectively. The state of equilibrium is never permanent, but the dynamic balance is being constantly readjusted under changing conditions of the environment.

The movement of the tree furnishes an example of the *negative* type of THERMONASTIC MOVEMENT. Parallel phenomena are found in floral organs, where, in the well-known instance of *Crocus*, the perianth leaves open outwards during rise of temperature and close inwards during the onset of cold. Looked at from above, the opening outwards during rise of temperature is a movement downwards, and therefore belongs to the *negative type.* In such cases the changed rate of growth by variation of temperature is the most important factor in the movement. It may be asked whether all thermonastic movements must necessarily belong to the *negative* type, where rise of temperature is attended by a movement downwards. I shall in my Paper on "Thermonastic Phenomena" show that there is also a *positive type* where rise of temperature induces an up-movement or of closure.

SUMMARY.

The 'Praying' Palm of Faridpur, growing at an inclination of about 60° to the vertical, exhibited a diurnal movement by which its head became erected in the morning and depressed towards the afternoon, the outspread leaves pressing against the ground.

The record of the diurnal movement showed that the head was erected to the highest position between 7 and 8 in the morning, after which there was a continuous fall which reached its climax at 3-15 P.M.; after this the movement was reversed and the maximum erection was again reached next morning.

This phenomenon is not unique, but is found exhibited, more or less, by all trees and their branches and leaves.

Diurnal records of temperature, and movement of the tree showed, that the two curves closely resembled each other. Rise of temperature was attended by a fall of the tree, and *vice versâ*.

The movement is brought about by the physiological action of temperature; it may be arrested by artificially induced physiological depression, and is permanently abolished at death.

The movement is primarily determined by the modifying influence of temperature on geotropic curvature. Rise of temperature is found to oppose or neutralise geotropic curvature, the fall of temperature inducing the opposite effect. The ever present tendency of upwards geotropic movement is opposed or helped by the effects of rise and fall of temperature respectively.

The movement of the "Praying" Palm is a thermonastic phenomenon. The tree, apparently so rigid, responds as a gigantic pulvinoid to the changes of its environment.

III.—ACTION OF STIMULUS ON VEGETABLE TISSUES

By

SIR J. C. BOSE,

Assisted by

NARENDRA NATH SEN GUPTA.

THE leaf of *Mimosa pudica* undergoes a rapid fall when subjected to any kind of shock. This plant has, therefore, been regarded as "sensitive," in contradistinction to ordinary plants which remain apparently immobile under external stimulus. I shall, however, show in course of this Paper that there is no justification in regarding ordinary plants as insensitive.

Let us first take any radial organ of a plant and subject it to an electric shock. It will be found that the organ undergoes a contraction in length in response to the stimulus. On the cessation of excitation the specimen gradually recovers its original length. Different organs of plant may be employed for the experiment, for example, the tendril of *Cucurbita*, the pistil of *Datura*, or the flower bud of *Crinum*. The shortening may be observed by means of a low power microscope. Greater importance is, however, attached to the detailed study of response and its time relations. The pull exerted by a delicate organ during its excitatory contraction is slight; hence arises the necessity of devising a very sensitive apparatus, which would give records magnified from ten to a hundred times.

RESPONSE RECORDERS.

The magnification of movement is produced by a light lever, the short arm of which is attached to the plant organ, the long arm tracing the record on a moving smoked plate of glass. The axis of the lever is supported by jewel bearings. The principal difficulty in obtaining accurate record of response of plant lies in the friction of contact of the recording point against the glass surface. This difficulty I have been able to overcome by providing a device of intermittent instead of continuous contact. For this, either the writer is made to vibrate to and fro, or the recording plate is made to oscillate backwards and forwards.

1. *The Resonant Recorder.*—In this the writing lever is made of a fine steel wire. One end of this wire is supported at the centre of a circular electromagnet; this latter is periodically magnetised by a coercing vibrator, which completes an electric circuit ten hundred, or two hundred times in a second. The writing lever is exactly tuned to the vibrating interrupter and is thus thrown into sympathetic vibration. Successive dots in the record thus measure time from 0·1 to 0·05 second. The employment of the Resonant Recorder enables us to measure extremely short periods of time for the determination of the latent period or the velocity of transmission of excitation.*

2. *The Magnetic Tapper.*—Measurement of very short intervals is not necessary in ordinary records of res-

* For detailed description *cf.* Bose.—"An Automatic Method for Investigation of Velocity of Transmission of Excitation in *Mimosa*."—Phil. Trans., B. vol. 204, (1913).

ponse. In this type of recorders, the circular magnet is therefore excited at longer intervals, from several seconds to several minutes; this is done by completion of the electric circuit at the required intervals, by means of a key operated by a clock.

3. *The Mechanical Tapper.*—In this, magnetic tapping is discarded in favour of mechanical tapping. The hinged writing lever is periodically pressed against the recording plate by a long arm, actuated by clock-work.

4. *The Oscillating Recorder.*—Here the plate itself is made to oscillate to-and-fro by eccentric worked by a clock. The frame carrying the plate moves on ball-bearings. The advantage of the Oscillating Recorder lies in the fact that a long lever, made of fine glass fibre, or of aluminium wire, may be employed for giving high magnification. A magnification of a hundred times may be easily obtained by making the short arm 2·5 mm. and the long arm 25 cm. in length.*

RESPONSE OF A RADIAL ORGAN.

Experiment 10.—As a typical example I shall describe the response of a straight tendril of *Passiflora.* A cut specimen was mounted with its lower end in water. Suitable electric connections were made for sending a feeble induction shock of short duration through the specimen. In this and all other records, unless contrary be stated, up-curve represents contractile movement. On application of stimulus of electric shock, an excitatory movement of contraction occurred which shortly reached its maximum; the apex-time was one minute and forty seconds, and recovery was completed after a further period of five minutes

* Bose—"Researches on Irritability of Plants," p. 279—Longmans, Green & Co.

(Fig. 10). Stronger shocks induce greater contraction

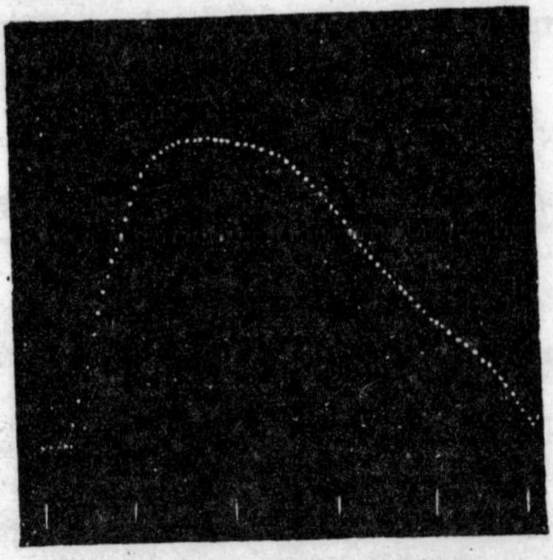

Fig. 10. Response of a straight tendril of *Passiflora* to electric shock. Successive dots at intervals of 5 seconds. The vertical lines below are at intervals of a minute. In this and in all following records (unless stated to the contrary) up-curve represents contraction, and down-curve expansion or recovery.

with prolongation of the period of recovery. The specimen was afterwards killed by application of poisonous solution of potassium cyanide; this brought about a permanent abolition of response. The experiment just described may be taken as typical of response of radial organs.

In a radial organ contraction takes place equally in all directions; it therefore shortens in length, there being no movement in a lateral plane. But if any agency renders one side less excitable than its opposite, diffuse stimulation will then induce greater contraction on the more excitable side which will therefore become concave.

RESPONSE OF AN ANISOTROPIC ORGAN.

Excessive stimulation is found to reduce the excitability of an organ. Under unilateral mechanical stimulation a tendril of *Passiflora* becomes hooked or coiled, the concave being the excited side. From what has been said, the unexcited convex side will relatively be the more excitable.

Experiment 11.—I took a specimen of hooked tendril, and excited it by an electric shock. The response was by the greater contraction of the more excitable convex side, on account of which the curved specimen tended to open out. The record of this response is seen in Fig. 11;

Fig. 11. Response of a hooked tendril of *Passiflora* to electric shock. Successive dots at intervals of 5 seconds.

the apex-time was nearly two minutes, and the recovery was completed in the further course of 15 minutes.

From the responses of organs rendered anisotropic by the differential action of the environment we pass to others which show certain amount of anatomical and physiological differentiation between their upper and lower sides. I find that many petioles of leaves show movement in response to stimulus. Many pulvini, generally regarded as insensitive, are also found to exhibit responsive movements.

RESPONSE OF THE PULVINUS OF *MIMOSA PUDICA*.

The most striking and familiar example of response is afforded by the main pulvinus of *Mimosa pudica* of which a record is given in Fig. 12 It is generally assumed

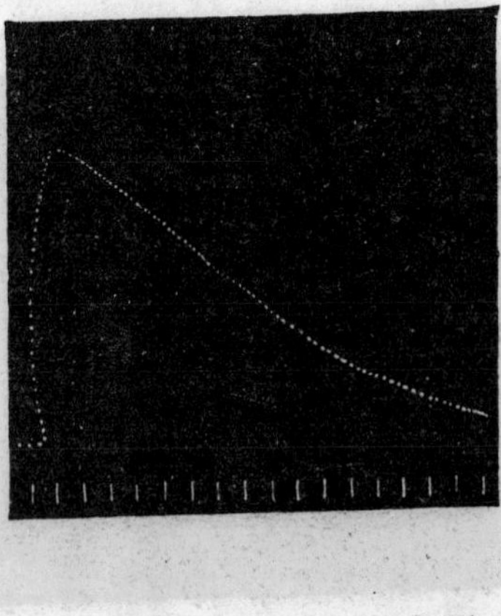

Fig. 12. Response of the main pulvinus of *Mimosa pudica*.

that sensibility is confined to the lower half of the organ. It will be shown in a subsequent Paper that this is not the case. The upper half of the pulvinus is also sensitive though in a feeble degree, its excitability being about 80 times less than that of the lower half. On diffuse stimulation the predominant contraction of the lower half causes the fall of the leaf, the antagonistic reaction of the upper half being, in practice, negligible. In order to avoid unnecessary repetition, I shall ignore the feeble antagonistic reaction of the less excitable half of the organ, and shall use the word 'contraction' for 'relatively greater contraction.'

It is interesting in this connection to refer to the response of the leaf of Water Mimosa (*Neptunia oleracea*). Here the reaction is very sluggish in comparison with that of *Mimosa pudica.* A tabular statement of contractile response of various radial, anisotropic and pulvinated organs will show a continuity in the contractile reaction ; the difference exhibited is a question of degree and not of kind.

TABLE I.—PERIODS OF MAXIMUM CONTRACTION AND OF RECOVERY OF DIFFERENT PLANTS.

Specimen.	Period of maximum contraction.	Period of recovery.
Radial organ :		
Tendril of *Passiflora*	100 seconds	4 minutes.
Anisotropic organ		
Hooked tendril of *Passiflora* ...	120 "	13 "
Pulvinated organ :		
Pulvinus of *Neptunia Oleracea* ...	180 "	57 "
Pulvinus of *Mimosa pudica* ...	3 "	16 "

As regards the excitatory fall of the leaf of *Mimosa pudica*, Pfeffer and Haberlandt are of opinion that this is due to the sudden diminution of turgor in the excited lower half of the pulvinus. The weight of the leaf, no longer supported by the distended lower cells, causes it to fall. This is accentuated by the expansion of the upper half of the pulvinus which is normally in a state of compression. According to this view the excitatory fall of the leaf is a passive, rather than an active, movement. I have, however, found that in determining the rapidity of the fall of *Mimosa* leaf the factors of expansive force of the upper half of the pulvinus and the weight of the leaf are negligible compared to the active force of contraction exerted by the lower half of the pulvinus (p. 87).

With regard to the fall of turgor, it is not definitely known whether excitation causes a sudden diminution in the osmotic strength of the cell-sap or an increase in the permeability of the ectoplast to the osmotic constituents of the cell. Pfeffer favours the former view, while others support the theory of variation of permeability.*

RESPONSE OF PULVINUS OF *MIMOSA* TO VARIATION OF TURGOR.

Whatever difference of opinion there may be in regard to the theories of osmotic and permeability variations, we have the indubitable fact of diminution of turgor and contractile fall of the pulvinus of *Mimosa* under excitation. The restoration of the original turgor brings about recovery and erection of the leaf. In connection with this the following experiments on responsive movements of the leaf under artificial variation of turgor will be found of interest:—

* With reference to the fall of *Mimosa* leaf Jost says: "When the pressure of the cell decreases we naturally assume this to be due to a decreasing *osmotic pressure* due to alterations in the permeability of the plasma, and an excretion of materials from the cell. It is a remarkable fact that plasmolytic research (Hilburg 1881) affords no evidence of any decrease in osmotic pressure. No complete insight into the mechanism of the stimulus movement in *Mimosa* has yet been obtained, although one thing is certain, that there is a decrease in the expansive power on the under side of the articulation."—Jost, "Plant Physiology"—English Translation, p. 515. Clarendon Press (1907). Blackman and Paine think that the loss of turgor on excitation "is probably due to the disappearance or inactivation of a considerable portion of the osmotic substances of the cells."—Annals of Botany, Vol. XXXII, No. CXXXV, Jan. 1918.

Effect of Increased Turgor: Experiment 12.—A young *Mimosa* plant was carefully transplanted and the root embedded in soil placed in a linen bag. This was held securely by a clamp, and one of the leaves of the plant attached to the recorder. Withholding of water for a day caused a general loss of turgor of the plant. A vessel full of water was now raised from below so that the linen bag containing the roots was now in water. The effect of increased turgor by suction of water by the roots became apparent by the *upward* movement of the leaf. The distance between the immersed portion of the plant and the leaf was 2 cm. and the up-movement of the leaf was indicated within 10 seconds of application of water (Fig. 13). The velocity with which the effect of

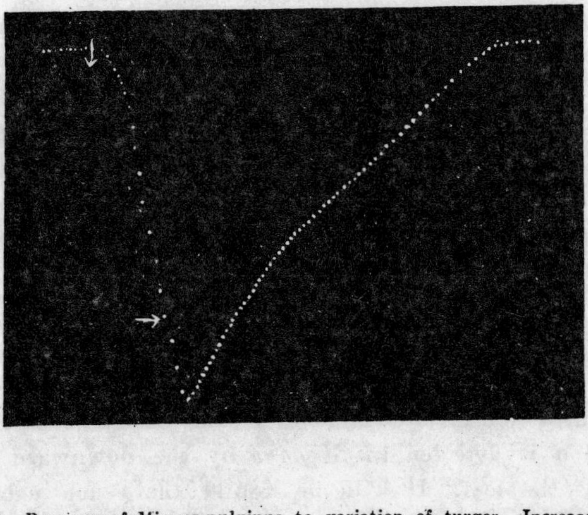

Fig. 13. Response of *Mimosa* pulvinus to variation of turgor. Increased turgor by application of water at point marked with vertical arrow induced erectile movement. Diminution of turgor by application of KNO_3 solution at the point marked with the horizontal arrow, brought about the fall of the leaf within 80 seconds. Successive dots at intervals of 5 seconds. (The down curve represents up-movement and *vice versâ*.)

increased turgor travelled was thus 2 mm. per second. The leaf exhibited increasing erection with absorption of water.

Effect of Diminution of Turgor: Experiment 13.—While the leaf in the above experiment was in process of erection, a quick change was made by substituting KNO_3 solution for the water of the vessel in which the roots were immersed. The plasmolytic withdrawal of water at the roots gave rise to a wave of diminished turgor, the effect of which became perceptible within 40 seconds by the movement of *fall* of the leaf. (Fig. 13.)

DIFFERENT MODES OF STIMULATION.

In *Mimosa* excitation is manifested by the contraction of the pulvinus and the consequent movement of the leaf. But in most plants, excitatory movement cannot be realized on account of the rigidity of the plant structure, the thickness of the cell-wall and the want of facility for escape of water from the excited cells. I shall show later how excitation may be detected in the absence of mechanical movement.

As regards stimulation of vegetable tissues, there are various agencies besides electric shock, which induce excitatory contraction; these agencies I shall designate as stimuli. Excitation is detected in *Mimosa* by the downward movement of the leaf. It will be found that such excitatory movement is caused by a mechanical blow, by a prick or a cut, by the application of certain chemical agents, by the action of electric current and by the action of strong light. The study of the action of these stimuli will be given in greater detail in subsequent Papers.

I shall give below a general classification of different stimuli which cause excitation in vegetable tissues.

Electric Stimulus.—Induction shock, condenser discharge, the make of kathode and the break of anode.

Mechanical Stimulus.—Mechanical blow, friction, prick or cut.

Chemical Stimulus.—Effect of certain acids, and of other chemical substances.

Thermal Stimulus.—Sudden variation of temperature; application of heated wire.

Radiation Stimulus.—Luminous radiation of the more refrangible portion of the spectrum; ultra-violet rays; thermal radiation in the infra-red region.

All these different forms of stimulus induce an excitatory contraction, a diminution of turgor, and a negative mechanical response or fall of a motile leaf.

SUMMARY.

A radial organ responds to stimulus by contraction in length; as all its flanks are equally excitable there is no lateral movement under diffuse stimulus.

Physiological anisotrophy is induced in an organ, originally radial and isotropic, by the unequal action of the environment on its different sides. Diffuse stimulus induces a greater contraction of the more excitable side.

In a curved tendril the concave side is less excitable than the convex. Diffuse stimulus tends to straighten the curved tendril.

In the pulvinus of *Mimosa pudica*, the lower half is eighty times more excitable than the upper, and the fall of the leaf is due to the predominant contraction of the more excitable lower half.

A diminution of turgor takes place in the excited cells. Restoration of turgor brings about recovery of the leaf to its normal erect position. Independent experiments show that the fall of the leaf may be brought about by an artificial diminution of turgor, and the erection of the leaf by an increase of turgor.

IV.—DIURNAL VARIATION OF MOTO-EXCITABILITY IN *MIMOSA*

BY

SIR J. C. BOSE.

SEVERAL phenomena of daily periodicity are known, but the relations between the recurrent external changes and the resulting periodic variations are more or less obscure. As an example of this may be cited the periodic variation of growth. Here the daily periodicity exhibited by a plant is not only different in varying seasons, but it also differs in diverse species of plants. The complexity of the problem is very great, for not only are the direct effects of the changing environment to be taken into consideration but also their unknown after-effects. Even in the case of direct effect, different factors, such as light, temperature, turgor, and so on, are undergoing independent variations; it may thus happen that their reactions may sometimes be concordant and at other times discordant. The nyctitropic movement of plants affords another example of daily periodicity. The fanciful name of 'sleep' is often given to the closure of the leaflets of certain plants at night. The question whether plants sleep or not may be put in the form of the definite inquiry: Is the plant equally excitable throughout day and night? If not, is there any definite period at which it practically loses its excitability? Is there, again, another period at which the plant wakes up, as it were, to a condition of maximum excitability?

In the course of my investigations on the irritability of *Mimosa pudica*, I became aware of the existence of such a daily periodicity; that is to say, the moto-excitability of the pulvinus was found to be markedly diminished or even completely abolished at a certain definite period of the day; at another equally definite period, the excitability was observed to have attained its climax. The observations on the periodic variation of excitability appeared at first to be extremely puzzling. It might be thought, for example, that light would prove to be favourable for moto-excitability; in actual experiment the results apparently contradicted such a supposition: for the excitability of the plant was found much higher in the evening than in the morning. Favourable temperature, again, might be regarded as an important factor for the enhancement of the moto-excitability; it was, nevertheless, found that though the excitatory response was only moderate at that period of night when the temperature was at its minimum, yet the excitability was altogether abolished at another period when the temperature was several degrees higher. The obscurities which surrounded the subject were only removed as a result of protracted investigation and comparison of continuous automatic records made by the plant itself during several months, beginning with winter and ending in summer.

The question whether a plant like *Mimosa* exhibits diurnal variation of excitability can be experimentally investigated by subjecting the plant at every hour of the day and night to a test-stimulus of uniform intensity, and obtaining the corresponding mechanical responses. Under these circumstances the amplitude of response at any time will serve as a measure of the excitability of the plant at the particular time. Any periodic fluctuation of response will then demonstrate the periodic character of variation of excitability.

The investigation thus resolves itself into :—

The successful construction of a Response Recorder which will automatically record the response of the plant to uniform periodic stimulation at all hours of day or night;

the study of the effects of various external conditions on excitability ;

the diurnal variation of excitability and its relation to the changes of external conditions.

I will first give a diagrammatic view of the different parts of the apparatus which I devised for this investigation. The leaf of *Mimosa* is attached to one arm of a light aluminium lever, L, by means of thread. At right angles to the lever is the writing index W, which traces on a smoked glass plate allowed to fall at a definite rate by clockwork the responsive movement of the leaf. Under a definite stimulus of electric shock the leaf falls down, pulling the lever L, and moving the writer towards the left. (Fig. 14.) The amplitude of the response-curve

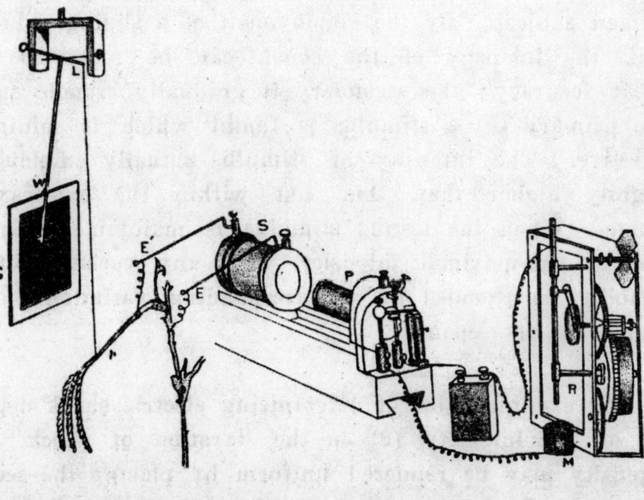

FIG. 14. Diagrammatic representation of the complete apparatus for determination of diurnal variation of excitability. Petiole of *Mimosa,* attached by thread to one arm of lever L ; writing index W traces on smoked glass plate G, the responsive fall and recovery of leaf. A, primary, and S, secondary, of induction coil. Exciting shock passes through the plant by electrodes E, E. A, accumulator. C, clockwork for regulating duration of tetanizing shock. Primary circuit of coil completed by plunging rod, V, dipping into cup of mercury M.

* *See* also Bose.—The Diurnal Variation of Moto-excitability in *Mimosa*—Annals of Botany, Oct. 1913.

measures the intensity of excitation. The leaf re-erects itself after a time, the corresponding record exhibiting recovery. A second stimulus is applied after a definite interval, say an hour, and the corresponding response shows whether the excitability of the plant has remained constant or undergone any variation.

UNIFORM PERIODIC STIMULATION.

Electric mode of excitation.—I find that one of the best methods of stimulating the plant is by means of tetanizing induction shock. The sensitiveness of *Mimosa* to electric stimulation is very great; the plant often responds to a shock which is quite imperceptible to a human subject. By the employment of a sliding induction coil, the intensity of the shock can be regulated with great accuracy; the secondary if gradually brought nearer the primary till a stimulus is found which is minimally effective. The intensity of stimulus actually employed is slightly higher than this, but within the sub-maximal range. When the testing stimulus is maintained constant and of sub-maximal intensity, then any variation of excitability is attended by a corresponding variation in the amplitude of response.

The exciting value of a tetanizing electric shock depends (*1*) on the intensity, (*2*) on the duration of shock. The intensity may be rendered uniform by placing the secondary at a fixed distance from the primary, and keeping the current in the primary circuit constant. The constancy of the current in primary circuit is secured by the employment of an accumulator or storage cell of definite electromotive force. It is far more difficult to secure the constant duration of the tetanizing shock in successive stimulations at intervals of, say, one hour during twenty-four hours. The duration of the induction shock given by the secondary coil depends on the length of time during which the primary circuit is completed in successive excitations. I have succeeded in overcoming the difficulty of securing uniformity of duration of shock by the employment of a special clockwork device.

The clockwork plunger.—The alarum clock can be so arranged that a wheel is suddenly released and allowed to complete one rapid revolution at intervals of, say, one hour. There is a fan-governor by which the speed of the revolution can be regulated and maintained constant. This will specially be the case when the alarum spring is long and fully wound. The succession of short releases twenty-four times during the day produce relatively little unwinding of the spring. On account of this and the presence of the fan-governor, the period of a single revolution of the wheel remains constant. By means of an eccentric the circular movement is converted into an up and down movement. The plungiug rod R thus dips into a cup of mercury M, for a definite short interval and is then lifted off. The duration of closure can be regulated by raising or lowering the cup of mercury. In practice the duration of tetanizing shock is about 0·2 second.

The same clock performs three functions. The axis which revolves once in twelve hours has attached to it a wheel, and round this is wound a thread which allows the recording glass plate to fall through six inches in the course of twenty-four hours. A spoke attached to the minute hand releases the alarum at regular and pre-determined intervals of time, say once in an hour. The plunging rod R, actuated by the eccentric, causes a tetanizing shock of uniform intensity and duration to be given to the plant at specified times.

Constancy of resistance in the secondary circuit.—In order that the testing electric stimulus shall remain uniform, another condition has to be fulfilled, namely, the maintenance of constancy of resistance in the secondary circuit, including the plant. Electric connections have to be made with the latter by means of cloth moistened with dilute salt solution; drying of the salt solution, however, gives rise to a variation of resistance in the electrolytic contact. This difficulty is overcome by making the electrolytic resistance negligible compared to the resistance offered by the plant. Thin and flexible spirals of silver tinsel attached to the electrodes E, E^1 are tied round the petiole and the stem, respectively. In order to secure better electric contact, a small strip of cloth moistened with dilute salt and glycerine is wound round the tinsel. As the resistance of contact is relatively small, and as drying is to a great extent retarded by glycerine, the total resistance of the secondary circuit undergoes practically no variation in the course of twenty-four hours. This will be seen from the following data. An experiment was commenced one day at 1 P.M., when the resistance offered by 8 cm. length of stem and 2 cm. length of petiole was found to be 1·5 million ohms. After twenty-four hours' record, the resistance was measured the next day and was found unchanged. The fact that the stimulus remains perfectly uniform will be quite apparent when the records given in the course of this paper are examined in detail.

THE RESPONSE RECORDER.

The amplitude of response affords, as we have seen, a measure of the excitability of the plant. In actual record friction of the writer against the glass surface becomes a source of error. This difficulty I have been able to overcome by the two independent devices, the Resonant Recorder and the Oscillating Recorder. In the former the writer is maintained by electric means in a state of continuous to and fro vibration, about ten times in a second. There is thus no continuous contact between the writer and the smoked glass surface, friction being thereby practically eliminated. The writer in this case taps a record, the successive dots occurring at intervals of 0·1 second. The responsive fall of the leaf is rapid, hence the successive dots in this part of the record are widely spaced; but the erection of the leaf during recovery takes place slowly, hence the recovery part of the curve appears continuous on account of the superposition of the successive dots. The advantage of the Resonant Recorder is that the curve exhibits both response and recovery. This apparatus is admirably suited for experiments which last for a few hours. There is, however, some drawback to its use in experiments which are continued for days together. This will be understood when we remember that for the maintenance of 10 vibrations of the writer in a second, 10 electric contacts have to be made; in other words, 36,000 intermittent electric currents have to be kept up per hour. This necessitates the employment of an electric accumulator having a very large capacity.

In the Oscillating Recorder the recording plate itself moves to and fro, making intermittent contact with the writer about once in a minute. The recording smoked glass plate is allowed to fall at a definite rate by the unwinding of a clock wheel. By an electromagnetic arrangement the holder of the smoked glass plate is made to oscillate to and fro, causing periodic contact with the writer.

The Oscillator is diagrammatically shown in Fig. 15.

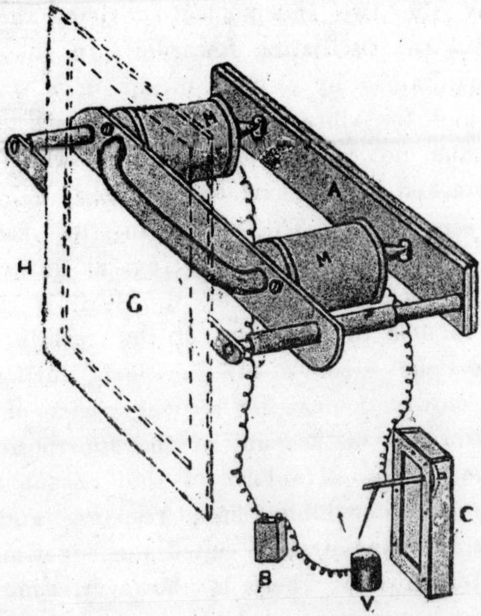

FIG. 15. The Oscillator. Electromagnet M, M′, periodically magnetized by completion of electric current by clockwork C. Periodic attraction of soft iron armature A moves attached glass plate G to left, making thereby electric contact with writer.

M, M′ are the two electromagnetic coils, the free ends of the horseshoe being pointed. Facing them are the conical holes of the soft iron armature A. This armature carries two rods which slide through hollow tubes. The distal ends of the rods support the holder H, carrying the smoked glass plate. Under normal conditions, the plate-holder is held by suitable springs, somewhat to the right of, and free from contact with, the writer. A clockwork C carries a rotating arm, which makes periodic contact with a pool of mercury contained in the vessel V, once in a minute. On the completion of the electromagnetic circuit, the armature A is attracted, the recording glass plate being thereby moved to the left making contact with the writer. The successive dots in the record thus take place at intervals of a minute. Only a moderate amount of electric current is thus consumed in maintaining the oscillation of the plate. A 4-volt storage cell of 20 amperes capacity is quite sufficient to work the apparatus for several days.

The responsive fall of the leaf of *Mimosa* is completed in the course of about two seconds. The leaf remains in the fallen or 'contracted' position for nearly fifteen seconds; it then begins to recover slowly. As the successive dots of the Oscillating Recorder are at intervals of a minute, the maximum fall of leaf is accomplished between two successive dots. The dotted response record here obtained exhibits the recovery from maximum fall under stimulation (*cf.* Fig. 23). The recovery of the leaf in one minute is less than one-tenth the total amplitude of the fall, and is proportionately the same in all the response records. Hence the successive amplitudes of response curves that are recorded at different hours of the day afford us measures of the relative variations of excitability of the plant at different times. This enables us to demonstrate the reality of diurnal variation of excitability. In my experimental investigations on the subject I have not been content to take my data from any particular method of obtaining response, but have employed both types of recorders, the Resonant and Oscillating. It will be shown that the results given by the different instruments are in complete agreement with each other.

EFFECTS OF EXTERNAL CONDITIONS ON EXCITABILITY.

Before giving the daily records of periodic variation of excitability, I will give my experimental results on the influence of various external conditions in modifying excitability. The conditions which are likely to affect excitability and induce periodicity are, first, the effects of light and darkness: under natural conditions the plant is subjected in the morning to the changing condition from darkness to light; then to the action of continued light during the day; and in the evening to the changing condition from light to darkness. A second periodic factor is the change in the condition of turgidity, which is at its maximum in the morning, as evidenced by the characteristic erect position of the petiole. Finally, the plant in the course of day and night is subjected to a great variation of temperature. I will now describe the effects of these various factors on excitability. It should be mentioned here that the experiments were carried out about the middle of the day, when the excitability, generally speaking, is found to remain constant.

EFFECTS OF LIGHT AND DARKNESS.

I have frequently noticed that a depression of excitability occurred when the sky was darkened by passing clouds. This is clearly seen in the above records obtained with the Resonant Recorder. Uniform sub-maximal

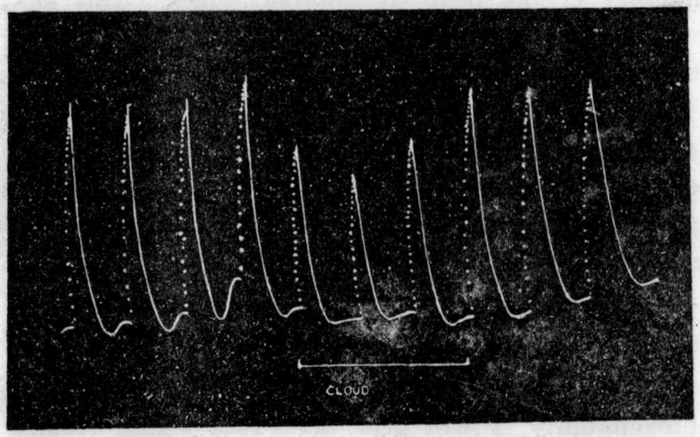

FIG. 16. Effect of cloud. Dotted up-curve indicates responsive fall, and continuous down-line exhibits slow recovery. First four responses normal; next three show depression due to diminution of light brought on by cloud, the duration of which is indicated by horizontal line below. Last three records show restoration of excitability brought on by clearing of sky. All records read from left to right

stimuli had been applied to a specimen of *Mimosa* at intervals of fifteen minutes. The dotted up-line represents the responsive fall, and the continuous down-line, the slow recovery. The first four are the normal uniform responses (Fig. 16). The next three show the depressing effect of relative darkness due to cloudy weather. The sky cleared after forty-five minutes, and we notice the consequent restoration of normal excitability.

Effect of sudden darkness and its continuation. Experiment 14.—In the next record (Fig. 17) is shown the

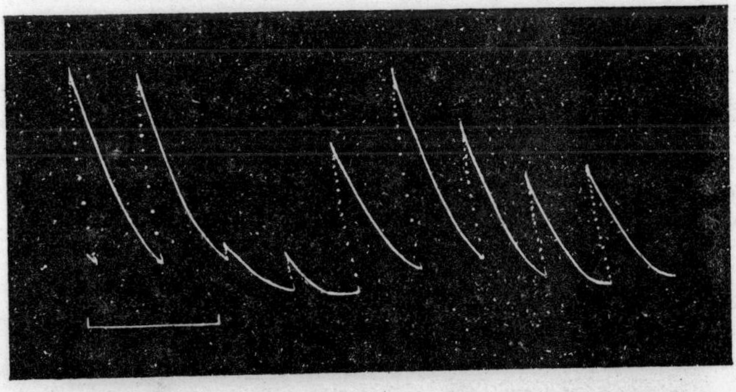

FIG. 17. Effect of sudden darkness. Plant subjected to sudden darkness beyond horizontal line seen below. First two responses normal. Note sudden depression of excitability, revival and final depression under continued darkness.

immediate and continued action of darkness. The first two are the normal uniform responses in light. By means of screens, the plant was next subjected to sudden darkness; this brought about a marked depression of excitability. Subjection to sudden darkness thus acts as a stimulus inducing a marked but transient fall of excitability. Under the continuous action of darkness, however, the excitability is at first restored and then undergoes a persistent depression.

Effect of transition from darkness to light : Experiment 15.—Here we have to deal first with the immediate effect of sudden transition, and then with the presistent effect of continuous light. In the record given in Fig. 18 the plant had been kept in the dark and the responses taken in the usual manner. It was then subjected to light ; the sudden change from darkness to light acted as a stimulus, inducing a transient depression of excitability. In this connection it is interesting to note that Godlewski found that in the phenomenon of growth, transition from darkness to light acted as a stimulus, causing a transient

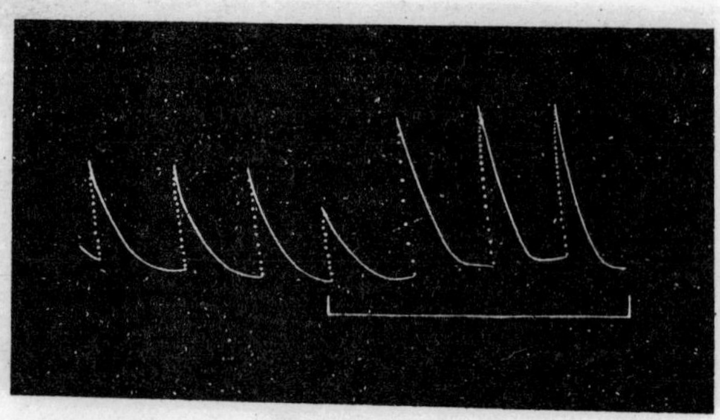

FIG. 18. Effect of change from darkness to light. The first three records are normal under darkness. Horizontal line below indicates exposure to light. Note preliminary depression followed by enhancement of excitability.

decrease in the normal rate. The effect of continued light on *Mimosa* is an enhancement of excitability.

EFFECT OF EXCESSIVE TURGOR.

I have often found that the moto-excitability is depressed under excessive turgor. Thus the over-turgid leaf of *Biophytum sensitivum* does not exhibit any mechanical response on rainy days.

Experiment 16.—The effect of excessive turgor on moto-excitability may be demonstrated in the case of *Mimosa*

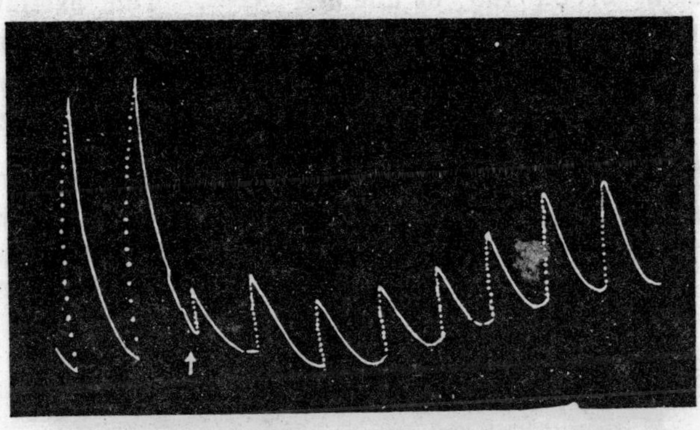

FIG. 19. Effect of enhanced turgor, artificailly induced. First two responses normal. Application of water, at arrow, induces depression of moto-excitability.

by allowing its main pulvinus to absorb water. The result is seen in the above record (Fig. 19), where water was applied on the pulvinus after the second response. It is seen how a depression of moto-excitability results from excessive turgor brought on by absorption of water. In such cases, however, the plant is found to accommodate itself to the abnormal condition and gradually regain its normal excitability in the course of one or two hours.

INFLUENCE OF TEMPERATURE.

The moto-excitability of the pulvinus of *Mimosa* is greatly modified under the influence of temperature. For the purpose of this investigation I enclosed the plant in a glass chamber, raising the temperature to the desired degree by means of electric heating. Responses to identical stimuli were then taken at different temperatures. It was found that the effect of heightened temperature, up to an optimum, was to enhance the amplitude of response.

Thus with a given specimen it was found that while at 22°C. the amplitude of response was 2·5 mm., it became 22 mm. at 27°C., and 52 mm. at 32°C. The excitability is enhanced under rising, and depressed under falling temperature. The moto-excitability of *Mimosa* is practically abolished at the minimum temperature of about 19°C.

Effect of lowering of temperature : Experiment 17.—A simple way of exhibiting the effect of lowering of temperature is by artificial cooling of the pulvinus. This cannot

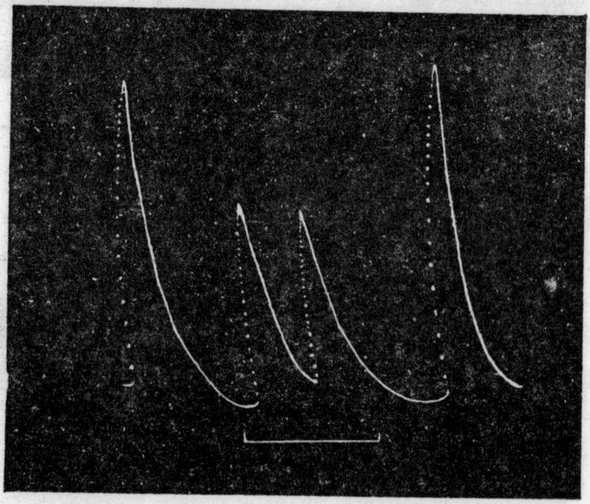

FIG. 20. Effect of moderate cooling during a period shown by horizontal line below. Moderate depression followed by quick restoration.

very well be done by application of a stream of cooled water, because, as we have seen, absorption of water by the pulvinus is attended by a loss of excitability : diluted glycerine has, however, no such drawback. This fluid at ordinary temperature was first applied on the pulvinus, and after an interval of half an hour records were taken in the usual manner. Cooled glycerine was then applied and the record taken once more ; the results are seen in

Figs. 20 and 21. In the former, the first response was normal at the temperature of the room, which was 32°C.; the next two exhibit depression of excitability under moderate cooling; the duration of application of moderately cooled glycerine is there indicated by the horizontal line below. On the cessation of application, the normal temperature was quickly restored, with the restoration of normal excitability.

In the next record (Fig. 21) is shown the effect of a more intense cold. It will be noticed that the first effect was a depression, and subsequently, a complete abolition of excitability. Thick dots in the record represent appli-

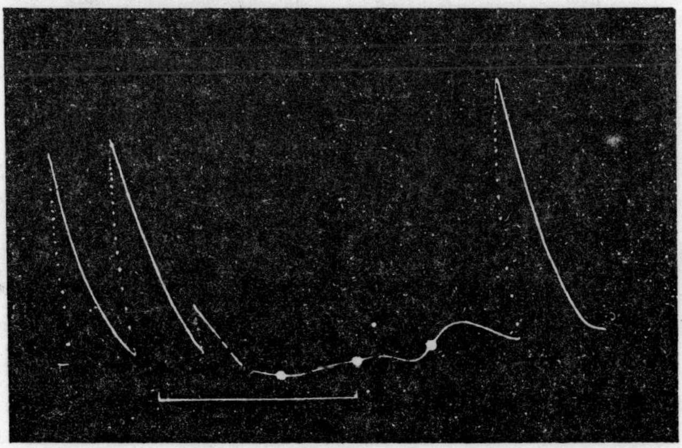

FIG. 21. Effect of application of more intense cold. Note sudden depression followed by abolition of excitability, also persistent after-effect.

cations of stimulus which proved ineffective. It will also be noticed that even on the cessation of cooling, and the return of the tissue to normal temperature, the induced abolition of excitability persisted as an after-effect for a considerable time. I have likewise found that the after-effect of cold in abolishing the conduction of excitation is also very persistent. These experiments show that owing to physiological inertia, the variations of excitability in the plant often lag considerably behind the external changes which induce them.

Effect of high temperature : Experiment 18.—It has been shown that the moto-excitability is enhanced by rising temperature ; there is, however, an optimum temperature above which the excitability undergoes a depression. This is seen in the following record (Fig. 22), where the normal response at 32°C. was depressed on raising the temperature to 42°C. ; the excitability was, however, gradually restored when the plant was allowed to regain the former temperature.

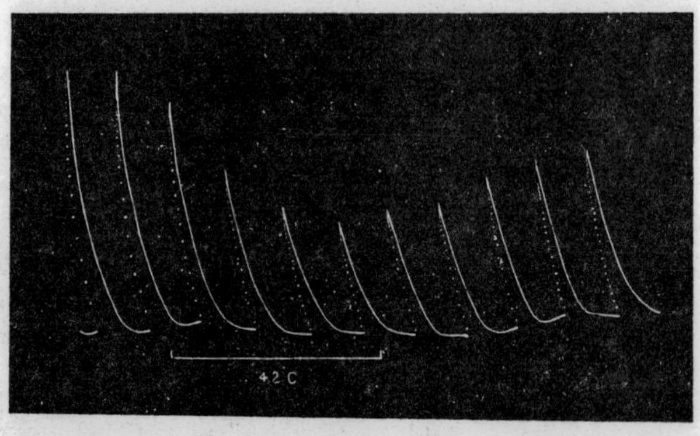

FIG. 22. Effect of temperature above optimum. Note depression of excitability induced by high temperature, and gradual restoration on return to normal.

I may now briefly recapitulate some of the important results : darkness depresses and light exalts the moto-excitability. Excessive turgor depresses motility. Still more marked is the effect of temperature. Lowering of temperature depresses and finally abolishes the moto-excitability : rise of temperature enhances it up to an optimum temperature, but beyond this point the excitability undergoes depression. *The change in excitability induced by the variation of external condition is not immediate; the induced effect, generally speaking, lags behind the inducing cause.*

DIURNAL VARIATION OF EXCITABILITY.

I will now give automatic records of responses taken once every hour for twenty-four hours. They prove conclusively the diurnal variation of excitability in *Mimosa.* After studying in detail the variations characteristic of particular times of the day, I will endeavour to correlate them with the effects brought on by the periodic changes of the environment.

Experiment 19.—As a typical example I will first give a record obtained in the month of February, that is, say, in spring. From this it will not be difficult to follow the variations which take place earlier in winter or later in summer.

The record given in Fig. 23 was commenced at 5 P.M. and continued to the same hour next day. The first thing noticeable is the periodic displacement of the base-line. This is due to the nyctitropic movements of the leaf. It should be remembered that the up-movement of the leaf is represented by down-curve, and *vice versâ.* After the maximum fall of the leaf, which in this case was attained at 9 P.M., there followed a reverse movement: the highest erection, indicative of maximum turgor, was reached at 6 A.M. The leaf then fell slowly and reached a middle position at noon. The extent of the nyctitropic movement varies in individual cases; in some it is slight, in others very large. The erectile movement began, as stated before, at about 9 P.M.; in some cases, however, it may occur as early as 6 P.M.

In following the characteristic variations of response,

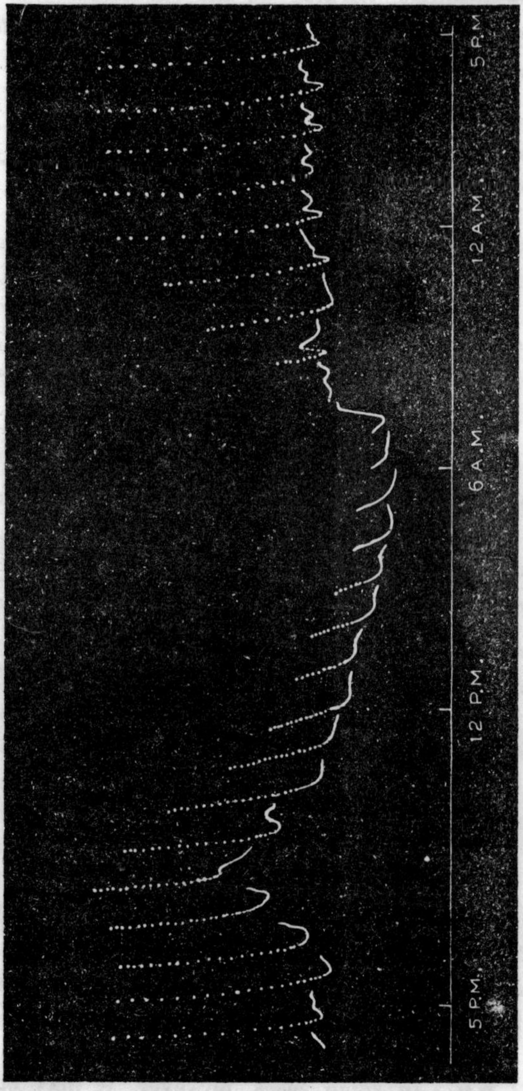

FIG. 23. Record for twenty-four hours, exhibiting diurnal variation of excitability (spring specimen). The displacement of base-line is due to nyctitropic movement.

occurring throughout the day, we find that while they

are practically uniform between the hours of 5 and 6 P.M., a continuous decline is manifested after setting in of darkness (7 P.M.); the fall of excitability continues even after sunrise (6-30 A.M.), response being practically abolished at 8 A.M. The excitability is then gradually restored in a staircase manner, the maximum being reached after 12 noon. After attaining this, the excitability remains more or less constant till the evening. It will be noticed that the amplitude of response at 5 P.M. on the second day was the same as the corresponding response on the previous day.

The results of this and numerous other records taken in spring may be summarized as :—

1. The maximum excitability of *Mimosa* is attained between 1 and 3 P.M., and remains constant for several hours. In connection with the constancy of response at this period, it should be remembered that when the response is at its maximum a slight increase of excitability cannot further enhance the amplitude of response.
2. The excitability, generally speaking, undergoes a continuous decline from evening to morning, the response being practically abolished at or about 8 A.M.
3. From 8 A.M. to 12 noon, the excitability is gradually enhanced in a staircase manner, till the maximum excitability is reached after 1 P.M.

I have obtained numerous records in support of these conclusions, some of which are reproduced in the following figures. In these cases responses to uniform stimuli at intervals of half an hour were taken at different parts of the day, the recorder employed being of the Resonant type.

Mid-day record: Experiment 20.—The record of daily periodicity previously given shows that the excitability reaches its maximum after 12 noon, and that it remains constant at the maximum value for several hours. This fact is fully borne out in the following record obtained with a different specimen (Fig. 24). The responses were taken here from noon to 3 P.M., once every half-hour.

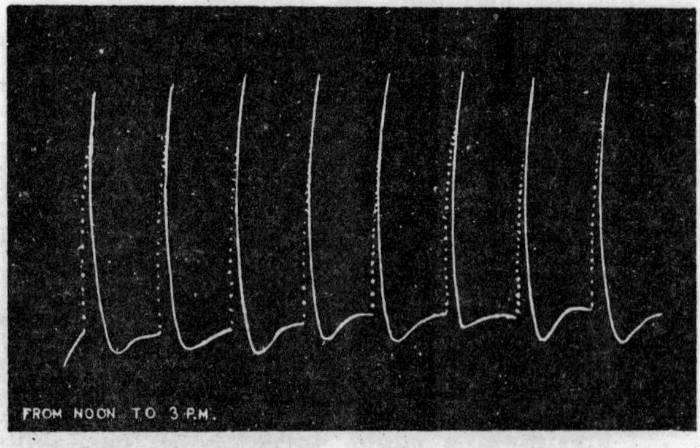

FIG. 24. Mid-day record from noon to 3 P.M. exhibiting uniform excitability. Responses taken once every half-hour.

Evening record: Experiment 21.—The record given in Fig. 23 shows that the amplitude of response falls continuously after 6 P.M. It might be thought that the diminished amplitude in the first part may be due to the natural nyctitropic fall of the leaf. The range of the pulvinar movement being limited, it is clear that the extent of the responsive fall must become smaller on account of the natural fall of the leaf during the first part of the night. That this is not the whole explanation of the decline of response in the evening will be clear from certain facts which I will presently adduce. It was stated that the leaf of *Mimosa* exhibits nyctitropic fal-

from 6 to 9 P.M., after which there is a reverse movement of erection. In certain specimens, however, the erectile movement commenced as early as 6 P.M. It is obvious that in these latter cases diminution of amplitude of response cannot be due to the reduction of the range of movement of the leaf. In Fig. 25 is given a series of

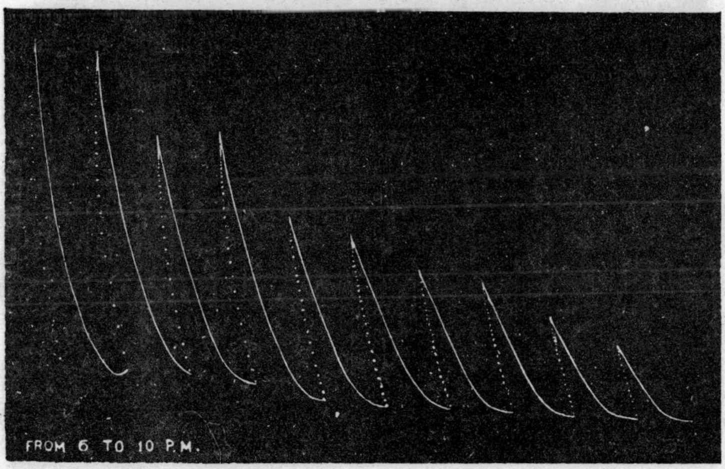

FIG. 25. Evening record from 6 to 10 P.M., showing gradual depression of excitability.

records from 6 to 10 P.M. obtained with a leaf in which erectile movement had commenced early in the evening. Though the full range of responsive movement was in this case available, yet the amplitude of successive responses is seen to undergo continuous diminution.

Record in the morning : Experiment 22.—The excitability is, as we have seen, nearly abolished about 8 A.M., after which there is a gradual restoration. This gradual enhancement of excitability to a maximum in the course of the forenoon is seen well illustrated in the above record (Fig. 26).

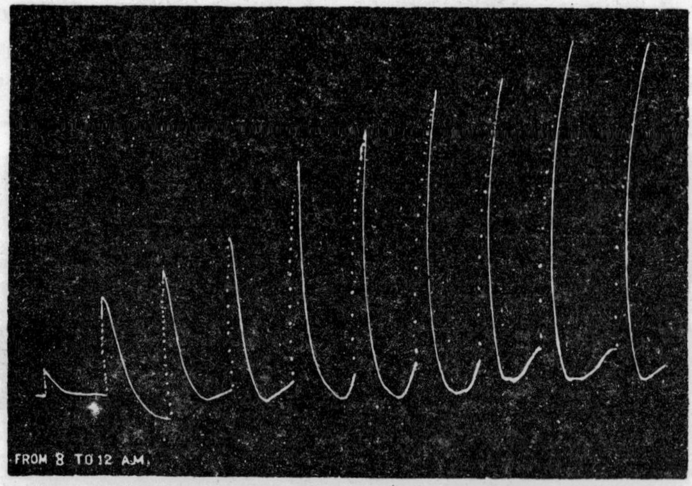

FIG. 26. Morning record from 8 A.M. to 12 noon, exhibiting gradual enhancement of excitability.

The record of daily periodicity given in Fig. 23 may be regarded as a typical example. Modifications may, however, be observed which are traceable to individual peculiarities. As an example of this, I give a record (Fig. 27) obtained with a specimen in which nyctitropic movement was very pronounced. The periodic variation of excitability exhibited here is practically the same as shown by other specimens. The interesting variation is in the character of the recovery from stimulus; the leaf was falling from 6 to 9 P.M.; owing to the shifting of the base-line upwards the recovery appears to be incomplete. After 9 P.M. the leaf was erected, at first slowly, then at a very rapid rate. The consequent fall of the base-line late at night is very abrupt; hence there is an apparent over-shooting in the line of recovery.

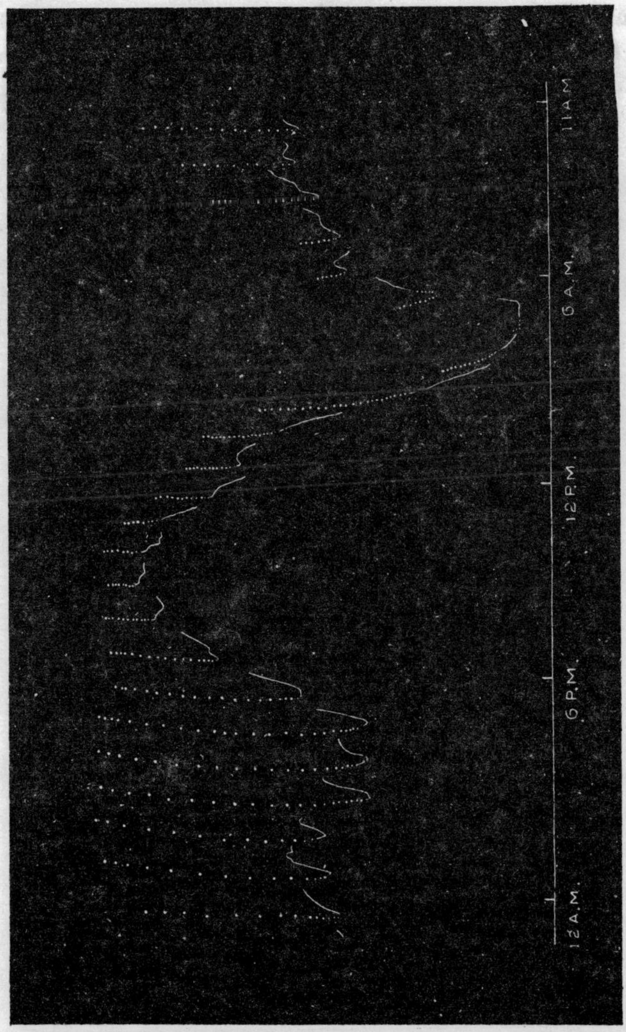

FIG. 27. Record of diurnal variation of excitability; it exhibits marked nyctitropic movement.

EFFECT OF TEMPERATURE ON VARIATION OF EXCITABILITY.

So far I have merely described the observed diurnal variation of excitability. We may next inquire whether there is any causal relation between the change of external conditions and the observed variation of excitability. It has been shown that the moto-excitability is greatly influenced by temperature. In order to find in what manner the diurnal variation of excitability was influenced by the daily variation of temperature, I took special care to secure by means of the thermograph a continuous record of temperature variations. The table which follows shows the relation between the hours of the day, temperature, and amplitude of response, in a typical case of diurnal variation of excitability.

TABLE II.—SHOWING THE RELATION BETWEEN HOUR OF THE DAY, TEMPERATURE AND EXCITABILITY. (SPRING SPECIMEN.)

Hours of day.	*Temperature*	*Amplitude of Response.*	*Hours of day.*	*Temperature*	*Amplitude of Response.*
5 p.m.	28° C.	28 mm.	5 a.m.	20° C.	5 mm.
6 „	25·5° „	28 „	6 „	20·5° „	4·2 „
7 „	24·5° „	27 „	7 „	21° „	3·5 „
8 „	23° „	23·5 „	8 „	22° „	2·5 „
9 „	22° „	21·5 „	9 „	24° „	0 „
10 „	21° „	18 „	10 „	26° „	6 „
11 „	20·5° „	15 „	11 „	26·5° „	15·5 „
12 „	20° „	13 „	12 „	28° „	22·5 „
1 a.m.	20° „	10 „	1 p.m.	28° „	26 „
2 „	20° „	8 „	2 „	28·5° „	28 „
3 „	20° „	7·5 „	3 „	28·5° „	28 „
4 „	19·5° „	6 „	4 „	29° „	28 „

From the data given in the table, two curves have been obtained. One of these shows the relation between the hours of the day and temperature; the other exhibits the relation between the hours of the day and the excitability as gauged by the amplitude of response (Fig. 28). It will

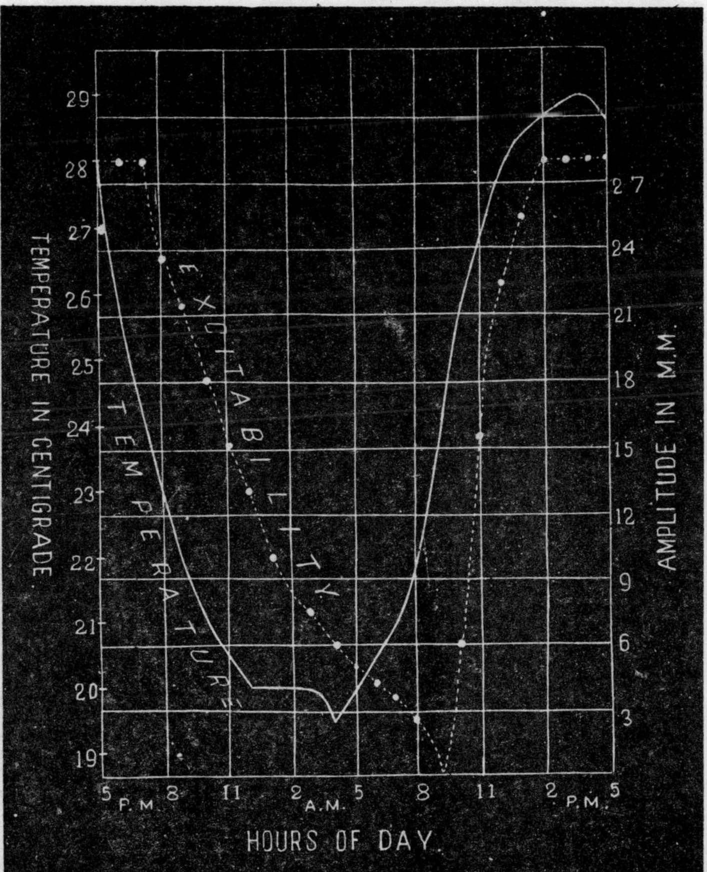

FIG. 28. The continuous curve shows the relation between the hour of the day and temperature. The dotted curve exhibits relation between the hour of the day and excitability.

be seen that there is, broadly speaking, a marked resemblance between the two curves, which demonstrate the predominant influence of temperature on diurnal variation of excitability.

EFFECT OF PHYSIOLOGICAL INERTIA.

It has been shown (page 59) that owing to physiological inertia, the change of excitability, generally speaking, lags behind the inducing cause. This fact finds striking illustration in the lag exhibited by the curve of excitability in reference to the temperature curve. The minimum temperature was attained at about 4 A.M., but the excitability was not reduced to a minimum till four hours later and again there is a marked fall of temperature after 5 P.M., but the excitability did not become depressed till two hours later.

There is again the factor of variation of light, the effect of which is not so great as that of temperature. The periods of maximum of light and temperature are, however, not coincident.

We may now discuss in greater detail the diurnal variation of excitability in *Mimosa*, taking the typical case, the record of which is given in Fig. 23. The temperature here is seen to remain almost constant, and at an optimum, from 1 to 5 P.M., the condition of light is also favourable. Hence the excitability is found to be constant, and at its maximum between these hours. The temperature begins to fall after 6 P.M., and there is, in addition, the depressing action of gathering darkness. Owing to the time-lag, the fall of excitability does not commence immediately at 6 P.M., but an hour afterwards, and continues till the next morning. During this period we have the cumulative effect of twelve hours' darkness and the increasing depression due to cold, the temperature minimum occurring at 4 A.M. On account of the combined effects of these various factors, and phenomenon of lag, the period of minimum excitability is in general reached about 8 A.M. In certain other cases this may occur earlier. After the attainment of this minimum, the excitability is gradually and continuously increased, under the action of light and of rising temperature, till the maximum is reached in the afternoon.

EFFECT OF SEASON.

It was said that temperature exerted a predominant influence in inducing variation of excitability. We may, therefore, expect that the diurnal period would be modified in a certain way according to the season. In winter the night temperature falls very low; hence the depression of excitability is correspondingly great, and results in the complete abolition of excitability. The after-effect of intense cold is seen in the condition of inexcitability persisting for a very long period in the morning. In summer the prevailing high temperature modifies the diurnal periodicity in a different manner. When the night is warm, the fall of excitability is slight. In the day, on the other hand, the temperature may rise above the optimum, bringing about a depression. In such a case the excitability in the earlier part of the evening may actually be greater than in the middle of the day. These modifications are shown in a very interesting way in the following record (Fig. 29) taken at the end of April. The temperature of Calcutta at this season often rises above 100°F. or 38°C. Table III also exhibits, in the case of the summer specimen, the relation between the hours of the day, temperature, and excitability.

An inspection of the record given in Fig. 29 shows that the amplitude of response was enhanced after 4 P.M. The temperature up to that time was unusually high (38 C.),

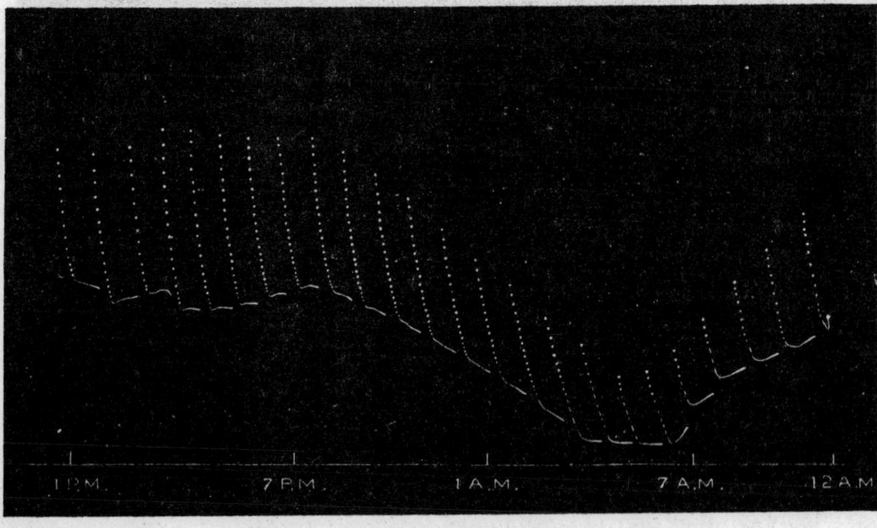

FIG. 29. Diurnal variation of excitability exhibited by summer specimen.

and there was in consequence a depression of excitability. After that hour there was a mitigation of heat, the temperature returning towards the optimum. Hence we find that the maximum excitability was attained between the hours 4 and 6 P.M. The minimum temperature at night was higher in the present case than that of the experiment carried out in February; in the former the minimum was 25·5°C., while in the latter it was 19·5 C. On account of this difference the night record in summer shows a fall of excitability which is far more gradual than that obtained in spring. The excitability is here not totally abolished in the morning, but reaches a minimum after 8 A.M.; the sensitiveness is then gradually enhanced in a staircase manner.

TABLE III—SHOWING THE RELATION BETWEEN HOURS OF THE DAY, TEMPERATURE, AND EXCITABILITY. (SUMMER SPECIMEN.)

Hours of day.	*Temperature.*	*Amplitude of Response.*	*Hours of day.*	*Temperature.*	*Amplitude of Response.*
1 p.m.	38° C.	22 mm.	1 a.m.	26° C.	21·5 mm.
2 „	38° „	23 „	2 „	26° „	20 „
3 „	38° „	24·5 „	3 „	25·5° „	18·5 „
4 „	37° „	28 „	4 „	25·5° „	17 „
5 „	35·5 „	29 „	5 „	25·5° „	16 „
6 „	33° „	27 „	6 „	26° „	15 „
7 „	31° „	26 „	7 „	27° „	14 „
8 „	30° „	26 „	8 „	29° „	13 „
9 „	29° „	25 „	9 „	30·5° „	11 „
10 „	27° „	24·5 „	10 „	33° „	16 „
11 „	27° „	24 „	11 „	35° „	1[illegible] „
12 „	26·5° „	22·5 „	12 „	37° „	21 „

SUMMARY.

The moto-excitability of *Mimosa* was gauged every hour of the day and night, by the amplitude of the response to a testing stimulus. This is effected by means of automatic devices which excite the plant periodically by an absolutely constant stimulus, and record the corresponding mechanical response.

From the record thus obtained, it was found that the excitability of the plant is not the same throughout the day, but undergoes a variation characteristically different at different times of the day. In a typical case in spring the excitability attained its maximum value after 1 P.M. and remained constant for several hours. There was then a continuous fall of excitability, the minimnm being reached at about eight in the morning. The plant at this time was practically insensitive. The moto-excitability was then gradually enhanced in a staircase manner till it again reached a maximum next afternoon.

The effect of sudden darkness was found to induce a transient depression, followed by revival of excitability. The effect of persistent darkness was to induce a depression.

Exposure to light from darkness caused a transient depression, followed by an enhancement of excitability.

Excessive turgor induced a diminished response.

Lowering of temperature induced a depression of excitability, culminating in an abolition of response. The after-effect of excessive cold was a prolonged depression of excitability.

Excitability was enhanced by rising temperature up to an optimum; above this point a depression was induced.

Owing to physiological inertia the change of excitability induced by variation of external condition lags behind the inducing cause.

The diurnal variation of excitability is primarily due to diurnal variation of temperature. The effect is modified in a minor degree by variation of light.

V.—RESPONSE OF PETIOLE-PULVINUS PREPARATION OF *MIMOSA PUDICA*

By

SIR J. C. BOSE,

Assisted by

SURENDRA CHANDRA DAS, M.A.

THE most suitable plant for researches on irritability of plants is *Mimosa pudica*, which can be obtained in all parts of the world. An impression unfortunataly prevails that the excitatory reaction of the plant can be obtained only in summer and under favourable circumstances; this has militated against its extensive use in physiological experiments, but the misgiving is without any foundation; for I found no difficulty in demonstrating even the most delicate experiments on *Mimosa* before the meeting of the American Association for the Advancement of Science held during Christmas of 1914. The prevailing outside temperature at the time was considerably below the freezing point. With foresight and care it should not be at all difficult to maintain in a hot-house a large number of these plants in a sensitive condition all the year round.

In order to remove the drawback connected with the supply of sufficient material, I commenced an investigation to find whether a detached leaf preparation could be made as effective for the study of irritability as the whole plant.

Here we have at the central end of the leaf the pulvinus, which acts as the contractile organ; the conducting strand in the interior of the petiole, on the other hand, is the vehicle for transmission of excitation. The problem to be solved is the rendering of an isolated petiole-and-pulvinus of *Mimosa* as efficient for researches on irritability as the nerve-and-muscle preparation of a frog. On the success of this attempt depended the practical opening out of an extended field of physiological investigation which would be unhampered by any scarcity of experimental material.

In connection with this it is well to note the surprising difference in vegetative growth as exhibited by plants grown in soil and in pots. A pot-specimen of *Mimosa* produces relatively few leaves, but one grown in the open ground is extremely luxuriant. As an instance in point, I may state that for the last five months I have taken from a plant grown in a field about 20 leaves a day for experiment, without making any impression on it. A large box containing soil would be practically as good as the open ground, and the slower rate of growth in a colder climate could be easily made up by planting half a dozen specimens. The protection of the plants from inclemencies of weather can be ensured by means of a glass cover with simple heat-regulation by electric lamps, in place of an expensive green-house.

Returning to the question of the employment of an isolated leaf, which I shall designate as a petiole-pulvinus preparation, instead of the entire plant, the first attempts which I made proved unsuccessful. The cut leaf kept in water would sometimes exhibit very feeble response, at other times all signs of excitability appeared to be totally abolished. It was impossible to attempt an investigation on the effect of changing environment on excitability when the normal sensitiveness itself underwent so capricious a change

These difficulties were ultimately overcome from knowledge derived through systematic investigation on the relative importance of the different parts of the motor apparatus, on the immediate and after-effect of section on the excitability of the leaf, and on the rate of decay of this excitability on isolation from the plant. The experience thus gained enabled me to secure long-continued and uniform sensibility under normal conditions. It was thus possible to study the physiological effects of changing external conditions by observing the responsive variation in the isolated petiole-pulvinus preparation. I propose to deal with the different aspects of the investigation in the following order :—

1. The effect of wound or section in modification of normal excitability.
2. The change of excitability after immersion in water.
3. Quantitative determination of the rate of decay of excitability in an isolated preparation.
4. Effect of amputation of the upper half of pulvinus.
5. Effect of removal of the lower half.
6. Influence of the weight of leaf on rapidity of responsive fall.
7. The action of chemical agents.
8. Effect of "fatigue" on response.
9. The influence of constant electric current on recovery.
10. The action of light and darkness on excitability.

The isolated petiole-pulvinus preparation is made by cutting out a portion of the stem bearing a single lateral leaf. The four diverging sub-petioles may also be cut off. In order to prevent rapid drying the specimen has to be kept in water. Preparations made in this way often appeared to have lost their sensibility. I was, however, able to trace this loss to two different factors: first, to the physiological depression due to injury caused by section, and, second, to the sudden increase of turgor brought on by excessive absorption of water. I shall now proceed to show that the loss of sensibility is not permanent, but is capable of restoration.

EFFECT OF WOUND OR SECTION IN MODIFICATION OF NORMAL EXCITABILITY.

In connection with the question of effect of injury, it is to be borne in mind that after each excitation the plant becomes temporarily irresponsive and that the excitability is fully restored after the completion of protoplasmic recovery. A cut or a section acts as a very intense stimulus, from the effect of which the recovery is very slow. If the stem be cut very near the leaf, the excitation of the pulvinus is very intense, and the consequent loss of excitability becomes more or less persistent. But if the stem be cut at a greater distance, the transmitted excitation is less intense, and the cut specimen recovers its excitability within a moderate time. I have also succeeded in reducing the excitatory depression by previously benumbing the tissue by physiological means. The isolated specimen can be made still more compact by cutting off the sub-petioles bearing the leaflets; the preparation now consists of a short length of stem of about 2 cm. and an equally short length of primary petiole, the motile pulvinus being at the junction of the two.

For the restoration of sensitiveness, and to meet working conditions, the lower end of the cut stem is mounted on a T-tube, with funnel-attachment and exit-tube, as shown in Fig. 30. The other two cut ends—of the stem and of the petiole—may be covered with moist cloth or may be closed with collodion flexile to prevent rapid evaporation and drying up of the specimen. A slight hydrostatic pressure maintains the specimen in a moderately turgid condition. A preparation thus made is

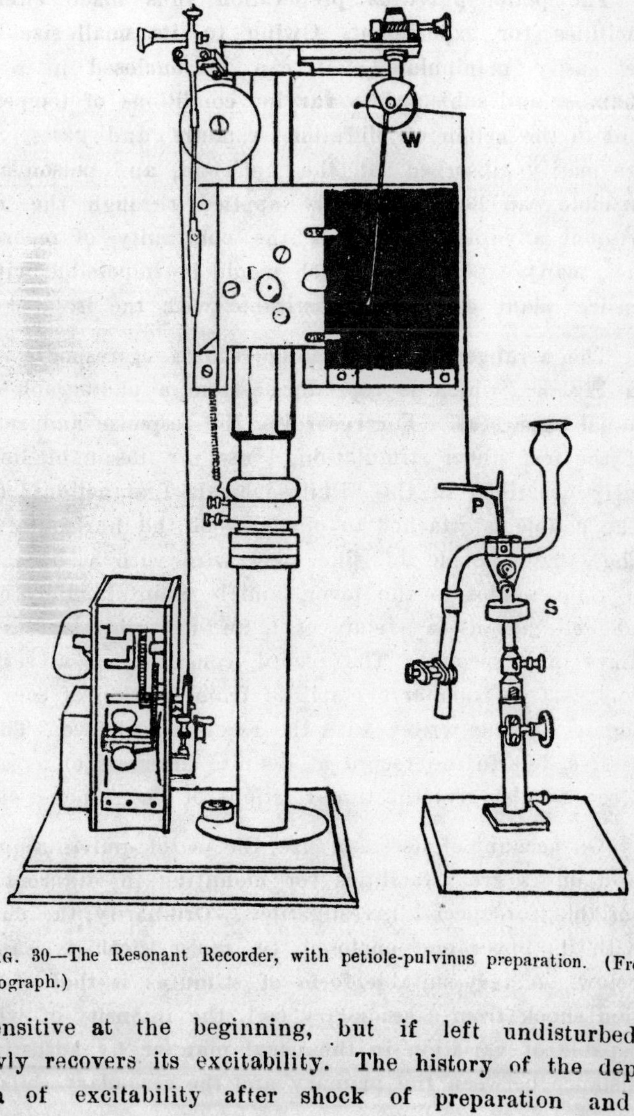

FIG. 30—The Resonant Recorder, with petiole-pulvinus preparation. (From a photograph.)

insensitive at the beginning, but if left undisturbed it slowly recovers its excitability. The history of the depression of excitability after shock of preparation and its gradual restoration is graphically illustrated by a series of records made by the plant (Fig. 31).

The petiole-pulvinus preparation thus made offers all facilities for experiment. Owing to its small size it can be easily manipulated; it can be enclosed in a small chamber and subjected to varying conditions of temperature and to the action of different vapours and gases. Drugs are easily absorbed at the cut end, and poison and its antidote can be successively applied through the funnel without any disturbance of the continuity of record. In fact, many experiments which would be impossible with the entire plant are quite practicable with the isolated leaf.

The arrangement for taking records of response is seen in Fig. 30, which is reproduced from a photograph of the actual apparatus. For recording the response and recovery of the leaf under stimulation, I use my Resonant Recorder fully described in the 'Philosophical Transactions' (1913). The petiole is attached to one arm of the horizontal lever. The writer, made of fine steel wire with a bent tip, is at right angles to the lever, and is maintained by electro-magnetic means in a state of to-and-fro vibration, say, ten times in a second. The record, consisting of a series of dots, is free from errors arising from friction of continuous contact of the writer with the recording surface. The successive dots in the record at definite intervals of a tenth of a second also give the time-relations of the response curve.

On account of its small size, the petiole-pulvinus preparation offers great facilities for mounting in different ways suitable for special investigations. Ordinarily, the cut stem with its lower end enclosed in moist cloth is supported below. A very suitable form of stimulus is that of induction shock from a secondary coil, the intensity of which is capable of variation in the usual manner by adjusting the distance between the primary and the secondary coils. The motile pulvinus, P, may be excited directly. For investigations on velocity of transmission of excitation, stimulus is applied on the petiole at some distance from the pulvinus, by means of suitable electrodes. Excitation is now transmitted along the intervening length of petiole, the conducting power of which will be found appropriately modified under the action of chemical and other agents. In this normal method of mounting, the more excitable lower half of the pulvinus is below ; excitatory reaction produces the fall of the petiole, gravity helping the movement. The preparation may, however, be mounted in the inverted position, with the more excitable lower half of the pulvinus facing upwards. The excitatory movement will now be the erection of the petiole, against gravity.

Under natural conditions the stem is fixed, and it is the petiole which moves under excitation. But a very interesting case presents itself when the petiole is fixed and the stem free. Here is presented the unusual spectacle of the plant or the stem "wagging" in response to excitation.

THE CHANGE OF EXCITABILITY AFTER IMMERSION IN WATER.

The isolated specimen can be kept alive for several days immersed in water. The excitability of the pulvinus, however, undergoes great depression, or even abolition, by the sudden change of turgor brought on by excessive absorption of water. The plant gradually accommodates itself to the changed condition, and the excitability is restored in a staircase manner from zero to a maximum.

In studying the action of a chemical solution on excitability, the solution may be applied through the cut end or directly on the pulvinus. The sudden variation of turgor, due to the liquid, always induces a depression, irrespective of the stimulating or the depressing action of the drug. The difficulty may be eliminated by previous long-continued application of water on the pulvinus and waiting till the attainment of uniform excitability which generally takes place in the course of about three hours. Subsequent application of a chemical solution gives rise to characteristic variation in the response.

QUANTITATIVE DETERMINATION OF THE RATE OF DECAY OF EXCITABILITY IN AN ISOLATED PREPARATION.

Variation of excitability after section : Experiment 23.— In order to test the history of the change of excitability resulting from the immediate and after-effect of section, I took an intact plant and fixed the upper half of the stem in a clamp. The response of a given leaf was now taken to the stimulus of an induction shock of 0·1 unit intensity, the unit chosen being that which causes a bare perception of shock in a human being. The specimen was vigorous and the response obtained was found to be a maximum. The stem bearing the leaf was cut at the moment marked in the record with a cross, and water was applied at the cut end. The effect of section was to cause the maximum fall of the leaf, with subsequent recovery. After this, successive responses to uniform stimuli at intervals of 15 minutes show, in (1) of Fig. 31, that a depression of excitability has been induced owing to the shock caused by section. In course of an hour, however, the excitability had been restored almost to its original value before the section. This was the case with a vigorous specimen, but with less vigorous ones a longer period of about three hours is required for restoration. In certain other cases the response after section exhibits alternate fatigue; that is to say, one response is large and the next feeble, and this alternation goes on for a length of time. The isolated specimen, generally speaking, attains a uniform sensibility after a few

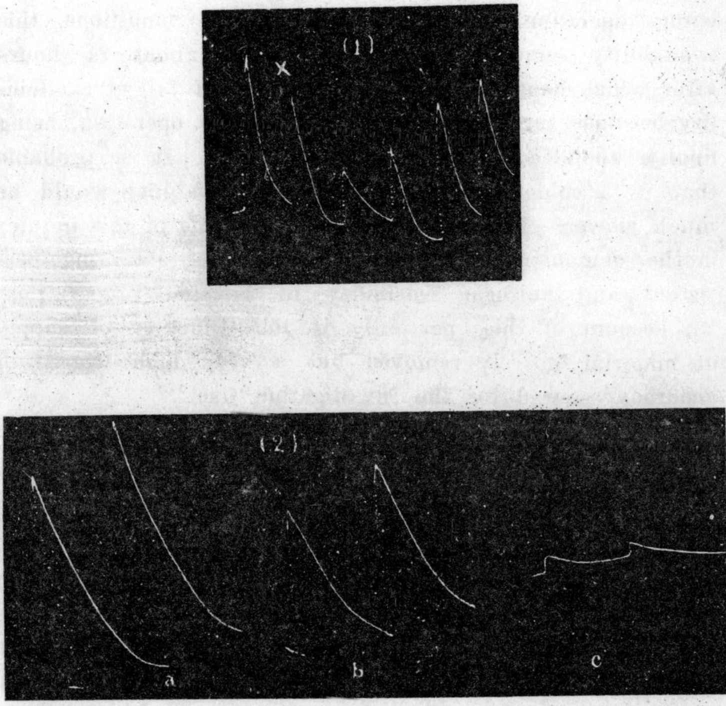

FIG. 31—Variation of excitability after section. (1) Immediate effect; (2) variation of excitability in a second specimen during 50 hours: (*a*) response 4 hours after section; (*b*) response after 24 hours; (*c*) after 49 hours. Up-line of record represents responsive fall of the leaf, down-line indicates recovery from excitation.

hours, which is maintained, with very slight decline under constant external conditions, for about 24 hours. On the third day the fall of excitability is very rapid, and the sensibility declines to zero in about 50 hours after isolation

[Fig. 31 (2)]. We may describe the whole cycle of change as follows: by the shock of operation the isolated preparation is rendered insensitive for nearly an hour, the excitability is then gradually restored almost to its normal value before operation. Under constant external conditions, this excitability remains fairly constant for about 24 hours after which depression sets in. The rate of fall of excitability becomes very rapid 40 hours after the operation, being finally abolished after the fiftieth hour. It is probable that in a colder climate the fall of excitability would be much slower. The most important outcome of this inquiry is the demonstration of the possibility of obtaining persistent and uniform sensibility in isolated preparations. On account of this, not only is the difficulty of supply of material entirely removed but a very high degree of accuracy secured for the investigation itself.

EFFECT OF AMPUTATION OF UPPER HALF OF PULVINUS.

Experiment 24.—The determination of the *rôle* played by different parts of the pulvinus in response and recovery is of much theoretical importance. Our knowledge on this subject is unfortunately very scanty. The generally accepted view is that on excitation "the actual downward curvature of the pulvinus is partly due to a contraction of the walls of the motor cells consequent upon the decrease of turgor, but is accentuated by expansion of the insensitive adaxial half of the pulvinus—which was strongly compressed in the unstimulated condition of the organ—and also by the weight of the leaf."* According to Pfeffer, after excitation of the organ, "the original condition of turgor is gradually reproduced in the lower half of the pulvinus, which expands, raising the leaf and producing compression of the

* Haberlandt, 'Physiological Plant Anatomy,' 1914, p. 570. English Translation, Macmillan & Co.

upper half of the pulvinus, which aids in the rapid curvature of the stimulated pulvinus."*

It was held, then, that the rapidity of the fall of leaf under stimulus is materially aided (1) by the expansion of the upper half of the pulvinus, which is normally in a state of compression, and (2) by the weight of the leaf. So much for theory. The experimental evidence available regarding the relative importance of the upper and lower halves of the pulvinus is not very conclusive. Lindsay attempted to decide the question by his amputation experiments. He showed that when the upper half was removed the leaf carried out the response, but rigor set in when the lower half was amputated. Pfeffer's experiments on the subject, however, contradicted the above results. He found that "after the upper half of the pulvinus was carefully removed, no movement was produced by stimulation, whereas when the lower half is absent a weakened power of movement is retained." Pfeffer, however, adds, "since the operation undoubtedly affects the irritability, it is impossible to determine from such experiments the exact part played by the active contraction of the lower half of the pulvinus."*

The cause of uncertainty in this investigation is twofold. First, it arises from the unknown change in irritability consequent on amputation; and, secondly, from absence of any quantitative standard by which the effect of selective amputation of the pulvinus may be measured. As regards the first, I have been able to reduce the depressing action caused by injury to a minimum by benumbing the tissue before operation, through local application of cold, and also allowing the shock-effect to disappear after a rest of several hours. As regards the physiological gauge of efficiency of the motor mechanism, such a measure is afforded by the

* PFEFFER—'Physiology of Plants,' vol. 3, pp. 75 and 76. English Translation, Clarendon Press.

relation between a definite testing stimulus and the resulting response with its time-relations, which is secured by my Resonant Recorder with the standardised electrical stimulator.

In carrying out this investigation I first took the record of normal response of an intact leaf on a fast moving plate. A second record, with the same stimulus, was taken after the removal of the upper half of the pulvinus, having taken the necessary precautions that have been described. Comparison of the two records (Fig. 32) shows that the only

FIG. 32.—Effect of amputation of upper half of pulvinus. Upper record gives normal response before amputation, and the lower, response after amputation (Successive dots at intervals of 0·1 sec.). Apex-time 1·1 sec., in both.

difference between them is in the exhibition of slight diminution of excitability due to operation. But, as regards the latent period and the quickness of attaining maximum fall, there is no difference between the two records before and after the amputation of the upper half. The upper part of the pulvinus is thus seen practically to have little influence in hastening the fall.

EFFECT OF REMOVAL OF THE LOWER HALF.

Experiment 25.—The shock-effect caused by the amputation of the lower half was found to be very great, and it required a long period of rest before the upper half regained its excitability. The excitatory reaction of the upper half is by contraction, and the response is, therefore, the lifting of the petiole. Thus, in an intact specimen, excitation causes antagonistic reactions of the two halves. But the sensibility of the upper half is very feeble and the rate of its contractile movement, relatively speaking, very slow. The record of the response of the upper half of the pulvinus, seen in Fig. 33, was taken with an Oscillating Recorder, where the successive dots are at intervals of 1 sec. : the

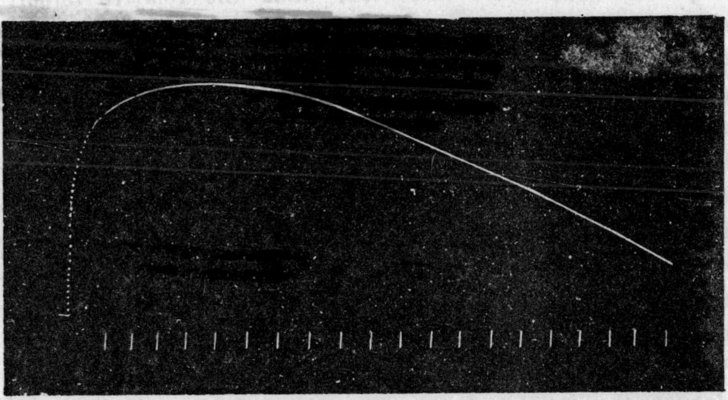

FIG. 33.—Response after amputation of lower half of pulvinus. (Successive dots at intervals of a second; vertical lines mark minutes.) Apex-time, 40 secs.

magnification employed was about five times greater than in recording the response of the lower half (Fig. 32). The intensity of stimulus to evoke response had also to be considerably increased. Taking into account the factors of magnification and the intensity of stimulus for effective response, the lower half I find to be about 80 times more sensitive than the upper. Thus, under feeble stimulus the upper half exerts practically no antagonistic reaction. The excitatory response of the upper half is also seen to be very sluggish.

INFLUENCE OF THE WEIGHT OF LEAF ON RAPIDITY OF RESPONSIVE FALL.

Experiment 26.—It is obvious that the mechanical moment exerted by the weight of the leaf must help its responsive fall under excitation. But the relative importance of the factors of active contraction of the lower half of the pulvinus and of the weight, in the rate of the responsive down-movement, still remains to be determined. A satisfactory way of solving the problem would lie in the study of the characteristics of response-records taken under three different conditions: (1) When the leaf is helped in its fall by its weight; (2) when the action of the weight is eliminated; and (3) when the fall has to be executed against an equivalent weight. An approximation to these conditions was made in the following manner. We may regard the mechanical moment to be principally due to the weight of the four sub-petioles applied at the end of the main petiole. In a given case these sub-petioles were cut off, and their weight found to be 0·5 grm. The main petiole was now attached to the right arm of the lever, and three successive records were taken: (1) With no weight attached to the petiole; (2) with 0·5 grm. attached to its end; and (3) with 0·5 grm. attached to the left arm of the lever at an equal distance from the fulcrum. In the first case, the fall due to the excitatory contraction will practically have little weight to help it; in the second case, it will be helped by a weight equivalent to those of the sub-petioles with their attached leaflets; and in the third case, the fall will be opposed by an equivalent weight. We find that in these three cases there is very little difference in the time taken by the leaf to complete the fall (Fig. 34).

It has been shown that the presence or absence of the

FIG. 34.—Effect of weight on rapidity of fall. N, without action of weight; W, with weight helping; and A, with weight opposing.

upper half of the pulvinus makes practically no difference in the period of fall; it is now seen that the weight exerts comparatively little effect. We are thus led to conclude that in determining the rapidity of fall, the factors of expansive force of the upper half of the pulvinus and the weight of the leaf are negligible compared to the active force of contraction exerted by the lower half of the pulvinus.

ACTION OF CHEMICAL AGENTS.

In connection with this subject it need hardly be said that the various experiments which I had previously carried out with the intact plant can also be repeated with the isolated preparation. I will only give here accounts of experiments which are entirely new.

The chemical solution may be applied directly to the pulvinus, or it may be absorbed through the cut end, the absorption being hastened by hydrostatic pressure. The normal record is taken after observing precautions which have already been mentioned. The reaction of a given chemical agent is demonstrated by the changed character of the record. The effect of the drug is found to depend not merely on its chemical nature, but also on the dose. There is another very important factor—that of the tonic condition of the tissue—which is found to modify the result. The influence of this will be realised from the account of an experiment to be given presently, where an

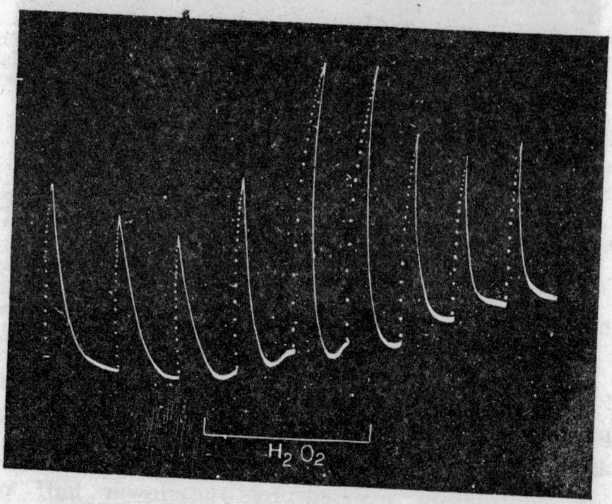

FIG. 35.—Stimulating action of hydrogen peroxide.

identical agent is shown to produce diametrically opposite effects on two specimens, one of which was in a normal, and the other in a sub-tonic, condition. The experiments described below relate to reactions of specimens in a normal condition.

Hydrogen Peroxide: Experiment 27.—This reagent in dilute solution exerts a stimulating action. Normal records, were taken after long-continued application of water on the pulvinus. The peroxide, as supplied by Messrs. Parke Davis & Co., was diluted to 1 per cent., and applied to the pulvinus; this gave rise to an enhancement of response. Re-application of water reduced the amplitude to the old normal value (Fig. 35).

Barium Chloride: Experiment 28.—The action of this agent is very characteristic, inducing great sluggishness in recovery. The preparation had been kept in 1-per cent. solution of this substance for two hours. After this the first response to a given test-stimulus was taken; the response was only moderate, and the recovery incomplete. The sluggishness was so great that the next stimulation, represented by a thick dot (Fig. 36), was ineffective. Tetanising electric shock at T, not only brought about

FIG. 36.—Incomplete recovery under the action of $BaCl_2$ and transient restoration under tetanisation at T.

response, but removed for the time being the induced sluggishness. This is seen in the next two records, which were taken under the old test-stimulus. There is now an enhanced response and a complete recovery. Beneficial effect of tetanisation disappeared, however, on the cessation of stimulus. This is seen in the next two records which were taken after two hours. The amplitude of response was not only diminished, but the recovery also was incomplete.

Antagonistic actions of Alkali and Acid: Experiment 29.—Alkali and acid are known to exert antagonistic actions on the spontaneous beat of the heart: dilute solution of NaOH arrests the beat of the heart in systolic contraction, while dilute lactic acid arrests the beat in diastolic expansion. I have found identical antagonistic reactions in the pulsating tissue of *Desmodium gyrans*, the

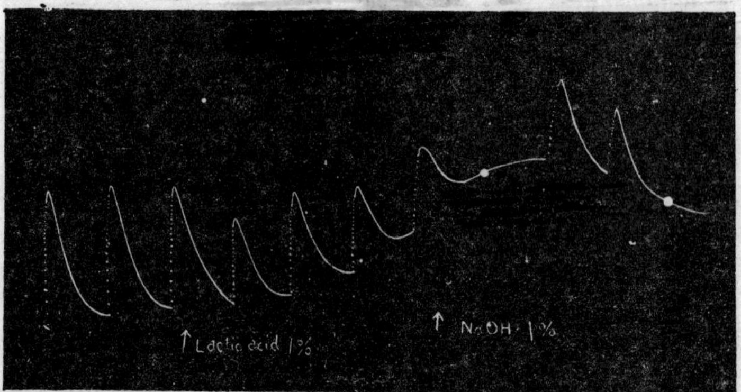

FIG. 37.—Antagonistic action of alkali and acid. Arrest of response in contraction under NaOH (↑), restoration and final arrest in expansion under actic acid (↑).

telegraph plant. It is very interesting to find that these agents also exert their characteristic effects on the response of *Mimosa* in a manner which is precisely the same. This is seen illustrated in Fig. 37, where the application of NaOH arrested the response in a contracted state; after this, the antagonistic effect of dilute lactic acid is seen first, in its power of restoring the excitability; its continued application, however, causes a second arrest, but this time in a state of relaxed expansion.

$CuSO_4$ Solution.—This agent acts as a poison, causing a gradual diminution of amplitude of response, culminating in actual arrest at death. Certain poisons, again, exhibit another striking symptom at the moment of death, an account of which will be given in a separate paper.

EFFECT OF "FATIGUE" ON RESPONSE.

With *Mimosa*, after each excitation the recovery becomes complete after a resting period of about 15 min. With this interval of rest the successive responses for a given stimulus are equal, and are at their maximum.

Experiment 30.—When the resting interval is diminished the recovery becomes incomplete, and there is a

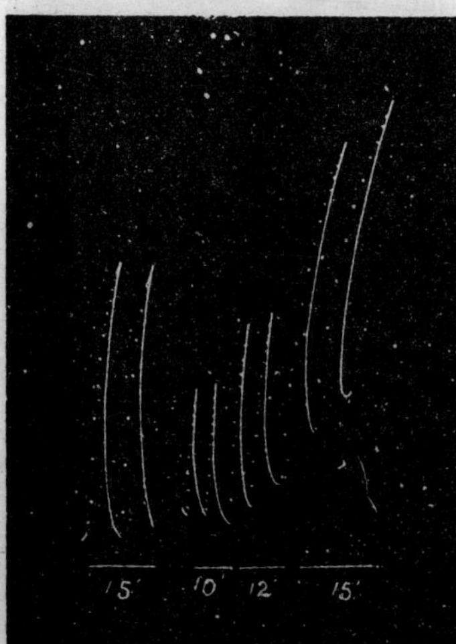

FIG. 38.—"Fatigue" induced by shortening intervening period of rest.

consequent diminution of amplitude of response. There is thus an increased fatigue with diminished period of rest. This is illustrated in Fig. 38, where the first two responses are at intervals of 15 min.; the resting interval was then reduced to 10 min., the response undergoing a marked diminution. Conversely, by increasing the resting interval, first to 12 and then to 15 min., the extent of fatigue was reduced and then abolished.

THE INFLUENCE OF CONSTANT ELECTRIC CURRENT ON RECOVERY.

Experiment 31.—From the above experiment it would appear that since the incompleteness of recovery induces fatigue, hastening of recovery would remove it. With this idea I tried various methods for quickening the recovery of the excited leaf. The application of a constant electric current was found to have the desired effect. Two electrodes for introduction of current were applied, one on the stem and the other on the petiole, at some distance from the pulvinus. In order to avoid the excitatory effect of sudden application, the applied current should be in-

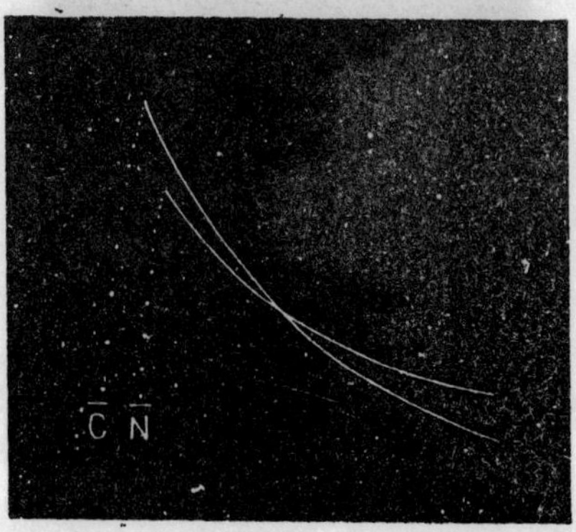

FIG. 39.—Action of constant current in removal of fatigue by hastening recovery; N, curve of response in fatigued specimen; C, after passage of current.

creased gradually; this was secured by means of a potentiometer slide. In my experiment a current having an intensity of 1·4 micro-ampere was found to be effective. Responses at intervals of 10 min., as we have seen, exhibit marked fatigue. Two responses were recorded on a fast-moving plate, N before, and C after, the application of the current. It will be seen (Fig. 39) how the application of current has, by hastening the recovery, enhanced the amplitude of response and brought about a diminution of fatigue. In connection with this, I may state that the tonic condition is, in general, improved as an after-effect of the passage of current. This is seen in some cases by a slight increase in excitability; in others, where the responses had been irregular, the previous passage of a current tends to make the responses more uniform.

ACTION OF LIGHT AND DARKNESS ON EXCITABILITY.

In taking continuous records of responses I was struck by the marked change of excitability exhibited by the intact plant under variation of light. Thus the appearance of a cloud was quickly followed by an induced depression, and its disappearance by an equally quick restoration of excitability. This may be explained on the theory that certain explosive chemical compounds are built up by the photo-synthetic processes in green leaves, and that the intensity of response depends on the presence of these compounds. But the building up of a chemical compound must necessarily be a slow process, and it is difficult on the above hypothesis to connect the rapid variation of excitability with the production of a chemical compound, or its cessation, concomitant with changes in the incident light.

Experiment 32.—In order to find out whether photo-synthesis had any effect on excitability, I placed an intact plant in a dark room and obtained from it a long series of responses under uniform test-stimulus. While this was being done the green leaflets were alternately subjected to strong light and to darkness, care being taken that the pulvinus was shaded all the time. The alternate action of light and darkness on leaflets induced no variation in the uniformity of response. This shows that the observed variation of excitability in *Mimosa* under the alternate action of light and darkness is not attributable to the photo-synthetic processes.

I next took a petiole-pulvinus preparation from which the sub-petioles bearing the leaflets had been cut off, and placed it in a room illuminated by diffused daylight. The normal responses were taken, the temperature of the room being 30° C. The room was darkened by pulling down the blinds, and records were continued in darkness. The temperature of the room remained unchanged at 30° C. It will be seen from records given in Fig. 40, that in

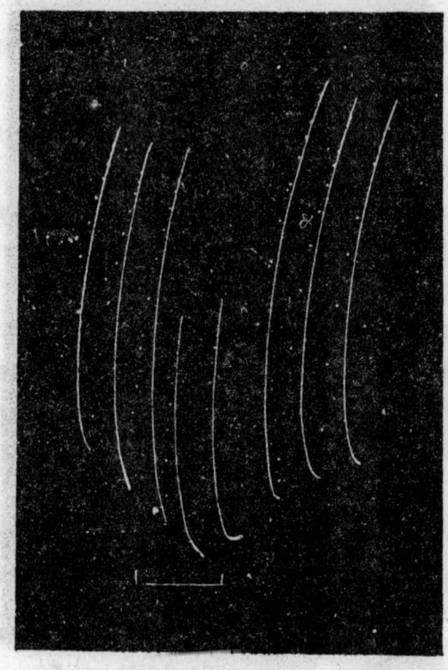

FIG. 40.—Stimulating action of light, and depressing action of darkness. Horizontal line below represents period of darkness.

darkness there is a great depression of excitability. Blinds were next pulled up and the records now obtained exhibit the normal excitability under light. The sky had by this time become brighter, and this accounts for the slight enhancement of excitability. This experiment proves conclusively that light has a direct stimulating action on the pulvinus, independent of photo-synthesis.*

SUMMARY.

On isolation of a petiole-pulvinus preparation, the shock of operation is found to paralyse its sensibility. After suitable mounting the excitability is restored, and remains practically uniform for nearly 24 hours. After this a depression sets in, the rate of fall of excitability becomes rapid 40 hours after the operation, sensibility being finally abolished after the fiftieth hour.

Experiments carried out on the effect of weight, and the influence of selective amputation of the upper and lower halves of the pulvinus, show that in determining the rapidity of fall of leaf, the assumed factors of the expansive force of the upper half of the pulvinus and the weight of the leaf are negligible compared to the force of active contraction exerted by the lower half of the pulvinus. The excitability of the lower half is eighty times greater than that of the upper.

Chemical agents induce characteristic changes in excitability. Hydrogen peroxide acts as a stimulant. Barium chloride renders the recovery incomplete; but tetanisation temporarily removes the induced sluggishness. Acids and

* *See* also Bose and Das—'Physiological Investigations with Petiole-Pulvinus preparations of *Mimosa pudica*.' Proc. Roy. Soc. B. Vol. 89, 1916.

alkalis induce antagonistic reactions, abolition of excitability with alkali taking place in a contracted, and with acid in an expanded condition of the pulvinus.

The responses exhibit fatigue when the period of rest is diminished. The passage of constant current is found to remove the fatigue.

Response is enhanced on exposure to light, and diminished in darkness. Light is shown to exert a direct stimulating action on the pulvinus, independent of photosynthesis.

VI.—ON CONDUCTION OF EXCITATION IN PLANTS

By

SIR J. C. BOSE.

THE plant *Mimosa* offers the best material for investigation on conduction of excitation. With regard to this question the prevailing opinion had been that in plants like *Mimosa*, there is merely a transmission of hydromechanical disturbance and no transmission of true excitation comparable with the animal nerve. I have, however, been able to show that the transmission in the plant is not a mechanical phenomenon, but a propagation of excitatory protoplasmic change. This has been proved by the arrest of conduction by the application of various physiological blocks. Thus local application of increasing cold retards, and finally abolishes the conducting power. The conducting tissue becomes paralysed for a time as an after-effect of application of cold; the lost conducting power may, however, be quickly restored by tetanising electric shocks. The conducting power of an animal nerve is arrested by an electrotonic block, the conductivity being restored on the cessation of the current. I have succeeded in inducing similar electrotonic block of conduction in *Mimosa*. Conductivity of a selective portion of petiole may also be permanently abolished by local action of poisonous solution of potassium cyanide.*

* BOSE—"An Automatic Method for the Investigation of Velocity of Transmission of Excitation in *Mimosa*." 'Phil. Trans.' 'B, Vol. 304 (1913) and also "Irritability of Plants." Longman's Green & Co. (1913), p. 132.

Having thus established the physiological character of the transmitted impulse in plants I shall now proceed to give some of the principal results of my earlier and recent investigations on the effects of various agencies on conduction of excitation in plants.

Apart from any question of hydro-mechanical transmission, it is important to distinguish two different modes of transmission of excitation. In a motile tissue contraction of a cell causes a physical deformation and stimulation of the neighbouring cell. Examples of this are furnished by the cardiac muscle of the animal, the pulvinus of *Mimosa*, and the stamen of *Berberis*. This mode of propagation may better be described as a *convection* of excitation.

The *conduction* of excitation, as in a nerve, is a different process of transmission of protoplasmic change. The conducting tissue in this case does not itself exhibit any visible change of form. In the plant the necessary condition for transmission of excitation to a distance is that the conducting tissue should be possessed of protoplasmic continuity in a greater or less degree. This condition is fulfilled by vascular bundles. There being greater facility of transmission along the bundles than across them, the velocity in the longitudinal direction is very much greater than in the transverse.

For accurate determination of velocity of transmission the testing stimulus should be quantitative and capable of repetition. Abnormal high velocity has been observed in *Mimosa* by applying crude and drastic methods of stimulation, by a transverse cut or a burn. This is apt to give rise to a very strong hydro-dynamic disturbance, which travelling with great speed, delivers a mechanical blow on the responding pulvinus. Such hydro-dynamic transmission is not the same as physiological conduction.

In the primary petiole of *Mimosa* the highest velocity under electric stimulation I find to be about 30 mm. per second. This velocity is considerably lower than the velocity in the nerve of higher animals, but higher than in the lower animals. As an example of the latter, mention may be made of the velocity of 10 mm. per second in the nerve of *Anodon* and 1 mm. per second in the nerve of *Eledone*.

PREFERENTIAL DIRECTION OF CONDUCTION.

Experiment 33.—The conduction of excitatory impulse takes place in both directions. This can be demonstrated by taking a petiole of *Biophytum sensitivum* or of *Averrhoa carambola*. These petioles are provided with a series of motile leaflets. Stimulation at the middle point of the petiole gives rise to two waves of excitation, one of which travels towards the central axis of the plant, and the other away from it. The centrifugal velocity is greater than the centripetal as will be seen from the following results:

Biophytum ...	Velocity in centrifugal direction ...	2·9 mm. per second.
	" centripetal " ...	2 mm. " "
Averrhoa ...	" centrifugal " ...	0·5 mm. " "
	" centripetal " ...	0·26 mm " "

EFFECT OF TEMPERATURE.

Variation of temperature has a marked effect on the velocity of transmission of excitation. Lowering of temperature diminishes the velocity, culminating in an arrest. Rise of temperature, on the other hand, enhances the velocity. This enhancement is considerable in specimens in which the normal velocity is low, but in plants in optimum condition, the velocity being already high, cannot be further enhanced. The following tabular statement gives results of effects of temperature on velocity of transmission in *Mimosa* and *Biophytum*:—

TABLE IV.—EFFECT OF TEMPERATURE ON VELOCITY OF TRANSMISSION.

Specimen.	Temperature.	Velocity.
Mimosa (winter specimen) ...	22°C	3·6 mm. per second.
	28°C	6·3 mm. " "
	31°C	9·0 mm. " "
Biophytum	30°C	3·7 mm. " "
	35°C	7·4 mm. " "
	37°C	9·1 mm. " "

EFFECT OF SEASON.

The velocity of transmission is very much lower in winter than in summer. In the petiole of *Mimosa*, the velocity in summer is as high as 30 mm. per second; in winter it is reduced to about 4 mm. The lowering of velocity in winter is partly due to the prevailing low temperature and also to the depressed state of physiological activity.

EFFECT OF AGE.

In a *Mimosa* plant, different leaves will be found of different age. Of these the youngest will be at the top. Lower down, we obtain a fully grown young leaf, and near the base, leaves which are very old. The investigation deals with the effect of age on the conducting power of the petiole.

Comparison of conducting power in different leaves: Experiment [illegible].—Selecting three leaves from the same plant we apply an identical electric stimulus at points 2 cm.

from the three responding pulvini. The electric connections are so made that the same tetanising shock is applied on the three petioles, very young, fully grown, and very old. The secondary coil is gradually pushed in till the leaves exhibit responsive fall. The fully grown leaf was the first to respond, the velocity of transmission being 23 mm. per second. The secondary coil had to be pushed nearer the primary through 6 cm. before excitation could be effectively transmitted through the young petiole; for the oldest leaf still stronger stimulus was necessary, since in this case the secondary had to be pushed through an additional distance of 4 cm. for effective transmission of excitation. I also determined the relative values of the minimal intensity of stimulus, effective in causing transmission of excitation in the three cases. Adopting as before the intensity of electric stimulus which causes bare perception in a human being as the unit, I find that the effective stimulus for a fully grown young petiole is 0·3 unit, while the very young required 2·5 units, and the very old 5 units. Hence it may be said that the conducting power of a very young is an eighth, and of the very old one-sixteenth of the conductivity of the fully grown young specimen.

It will thus be seen that the conducting power of a very young petiole is feebler than in a fully grown specimen. The conducting tissue, it is true, is present, but the power of conduction has not become fully developed. This power is, as we shall see later, conferred by the stimulus of the environment. In a very old specimen the diminution of conducting power is due to the general physiological decline.

EFFECT OF DESICCATION ON CONDUCTING TISSUES.

I have already shown that transmission in the plant is a process fundamentally similar to that taking place in the animal nerve; it has also been shown that the effects of various physical and chemical agents are the same in the conducting tissues of plant and of animal.

Effect of application of glycerine: Experiment 35.—It is known that desiccation, generally speaking, enhances the excitability of the animal nerve. As glycerine, by absorption of water, causes partial desiccation, I tried its effect on conduction of excitation in the petiole of *Mimosa.* Enhancement of conducting power may be exhibited in two ways: first, by an increase of velocity of transmission; and, secondly, by an enhancement of the intensity of the transmitted excitation, which would give rise to a greater amplitude of response of the motile indicator. In Fig. 41 are given two records, N, before, and the other after

FIG. 41—Action of glycerine in enhancing the speed and intensity of transmitted excitation. Stimulus applied at the vertical line. Successive dots in record are at intervals of 0·1 sec.

the application of glycerine on a length of petiole through which excitation was being transmitted. The time-records demonstrate conclusively the enhanced rate of transmission after the application of glycerine. The increased intensity of transmitted excitation is also seen in the enhanced amplitude of response seen in the more erect curve in the upper record.

INFLUENCE OF TONIC CONDITION ON CONDUCTIVITY.

Different specimens of *Mimosa* are found to exhibit differences in physiological vigour. Some are in an optimum condition, others in an unfavourable or sub-tonic condition. I shall now describe certain characteristic differences of conductivity exhibited by tissues in different conditions.

Effect of intensity of stimulus on velocity of transmission.—In a specimen at optimum condition, the velocity remains constant under varying intensities of stimulus. Thus the velocity of transmission in a specimen was determined under a stimulus intensity of 0·5 unit; the next determination was made with a stimulus of four times the previous intensity, *i.e.*, 2 units. In both these cases the velocity remained constant. But when the specimen is in a sub-tonic condition, the velocity is found to increase with the intensity of the stimulus. Thus the velocity of conduction of a specimen of *Mimosa* in a sub-tonic condition was found to be 5·9 mm. per second under a stimulus of 0·5 unit; with the intensity raised to 2·5 units, the velocity was enhanced to 8·3 mm. per second.

After-effect of stimulus.—In experimenting with a particular specimen of *Mimosa* I found that on account of its sub-tonic condition, the conducting power the petiole was practically absent. Previous stimulation was, however, found to confer the power of conduction as an after-effect. It is thus seen that stimulus canalises a path for conduction.

The effect of excessive stimulus in a specimen in an optimum condition is to induce a temporary depression of conductivity; the effect of strong stimulus on a sub-tonic specimen is precisely the opposite, namely, an enhancement of conductivity. I give below accounts of two typical experiments carried out with petiole-pulvinus preparation of *Mimosa*. Excessive stimulation in these cases was caused by injury.

Action of Injury on Normal Specimens: Experiment 36.—A cut stem with entire leaf was taken, and stimulus applied at a distance of 15 mm. from the pulvinus. From the normal record (1) in Fig. 42 the velocity of transmission was

FIG. 42.—Effect of injury, depressing rate of conduction in normal specimen; (1) record before, and (2) after injury. (Dot-intervals, 0·1 sec.).

found to be 18·7 mm. per sec. The end of the petiole beyond the point of application of the testing stimulus was now cut off, and record of velocity of transmission taken once more. It will be seen from record (2) that the excessive stimulus caused by injury had induced a depression in the conducting power, the velocity being reduced to 10·7 mm. per sec. Excessive stimulation of normal specimens is thus seen to depress temporarily the conducting power.

Action of Injury on Sub-tonic Specimens: Experiment 37.—I will now describe a very interesting experiment which shows how an identical agent may, on account of difference in the tonic condition of the tissue, give rise to diametrically opposite effects. In demonstrating this, I took a specimen in a sub-tonic condition, in which the conducting power of the tissue was so far below par, that the test-stimulus applied at a distance of 15 mm. failed to be transmitted (Fig. 43). The end of the petiole at a distance

FIG. 43.—Effect of injury in enhancing the conducting power of a sub-normal specimen; (1) Ineffective transmission becoming effective at (2) after section; (3) decline after half an hour, and (4) increased conductivity after a fresh cut.

of 1 cm. beyond the point of application of test-stimulus was now cut off. The after-effect of this injury was found to enhance the conducting power so that the stimulus previously arrested was now effectively transmitted, the velocity being 25 mm. per sec. This enhanced conducting power began slowly to decline, and after half an hour the velocity had declined to 4·1 mm. per sec. The end of the petiole was cut once more, and the effect of injury was again found to enhance the conducting power, the velocity of transmission being restored to 25 mm. per sec.

SUMMARY.

There are two different types of propagation of excitation: by convection, and by conduction. In the former the excited cell undergoes deformation and causes mechanical stimulation of the next; example of this type is seen in the stamen of *Berberis*. The conduction of excitation consists, on the other hand, of propagation of excitatory protoplasmic change. The tranmission in the petiole of *Mimosa* is a phenomenon of conduction.

This conduction takes place along vascular elements. The conductivity is very much greater in the longitudinal than in the transverse direction.

Rise of temperature enhances, and fall of temperature lowers, the rate of conduction. Excitation is transmitted in both directions; the centrifugal velocity is greater than the centripetal.

Dessication of conducting tissue by glycerine enhances the conducting power. Local application of cold depresses or arrests the conduction. Application of poison permanently abolishes the power of conduction.

Conductivity is modified by the effect of season, being higher in summer than in winter.

The power of conduction is also modified by age. In young specimens the conducting power is low, the conductivity is at its maximum in fully grown organs; but a decline of conductivity sets in with age.

The tonic condition of a tissue has an influence on conductivity. In an optimum condition, the velocity is the same for feeble or strong stimulus. Excessive stimulation induces a temporary depression of the conducting power.

The effects are different in a sub-tonic tissue: velocity of transmission increases with intensity of stimulus; after-effect of stimulus is to initiate or enhance the conducting power. The conducting path is canalised by stimulus.

VII.—ON ELECTRIC CONTROL OF EXCITATORY IMPULSE

By

SIR J. C. BOSE.

I HAVE in my previous works * described investigations on the conduction of excitation in *Mimosa pudica*. It was there shown that the various characteristics of the propagation of excitation in the conducting tissue of the plant are in every way similar to those in the animal nerve. Hence it appeared probable that any newly found phenomenon in the one case was likely to lead to discovery of a similar phenomenon in the other.

As the transmission of excitation is a phenomenon of propagation of molecular disturbance in the conducting vehicle, it appeared that the excitatory impulse could be controlled by inducing in the conducting tissue two opposite 'molecular dispositions', using that term in the widest sense. The possibility of accomplishing this by the directive action of an electric current had attracted my attention for many years.

METHOD OF CONDUCTIVITY BALANCE.

I have previously carried out an electric method of investigation, dealing with the influence of electric current on conductivity. The method of Conductivity Balance which I devised for this purpose† was found very suitable. Isolated conducting tissues of certain plants were found to exhibit

* BOSE—"Comparative Electro-Physiology" (1907). Longmans, Green and Co.

† *Ibid*, p. 478

transmitted effect of excitatory electric change of galvanometric negativity, which, at the favourable season of the year was of sufficient intensity to be recorded by a sensitive galvanometer. A long strand of the conducting tissue was taken and two electric connections were made with a galvanometer, a few centimetres from the free ends. Thermal stimulus was applied at the middle, when two excitatory waves with their concomitant electric changes were transmitted outwards. By suitably moving the point of application of stimulus nearer or further away from one of the two electric contacts, an exact balance was obtained. This was the case when the resultant galvanometer deflection was reduced to zero. If now an electrical current be sent along the length of the conducting tissue, the two excitatory waves sent outwards from the central stimulated point will encounter the electric current in different ways; one of the excitatory waves will travel with, and the other against the direction of the current. If the power of transmitting excitation is modified by the direction of an electric current then the magnitudes of transmitted excitations will be different in the two cases, with the result of the upsetting of the Conductivity Balance. From the results of experiments carried out by this method on the effect of feeble current on conductivity, the conclusion was arrived at that *excitation is better conducted against the direction of the current than with it.* In other words, the influence of an electric current is to confer a preferential or selective direction of conductivity for excitation, the tissue becoming a better conductor in an electric up-hill direction compared with a down-hill.

The results were so unexpected that I have for long been desirous of testing the validity of this conclusion by independent method of inquiry. I shall presently give full account of the perfected method, and the various difficulties which had to be overcome to render it practical. Before doing his I shall describe a simple method which I have devised for demonstrating the principal results.

CONTROL OF TRANSMITTED EXCITATION IN *AVERRHOA BILIMBI.*

The petiole of *Averrhoa bilimbi* has a large number of paired leaflets, which, on excitation, undergo downward closure. Feeble stimulus is applied at a point in the petiole, and the transmission of excitation is visibly manifested by the serial fall of the leaflets. The distance to which the excitation reaches is a measure of normal power of conduction. Any variation of conductivity, by the passage of an electric current in one direction or the other is detected by the enhancement or diminution of the distance through which excitation is transmitted. I shall describe the special precautions to be taken in carrying out this investigation.

Electric stimulus of induction shock of definite intensity and duration is supplied at the middle of the petiole at EE′ (Fig. 44). The leaflets to the left of E, are not necessary

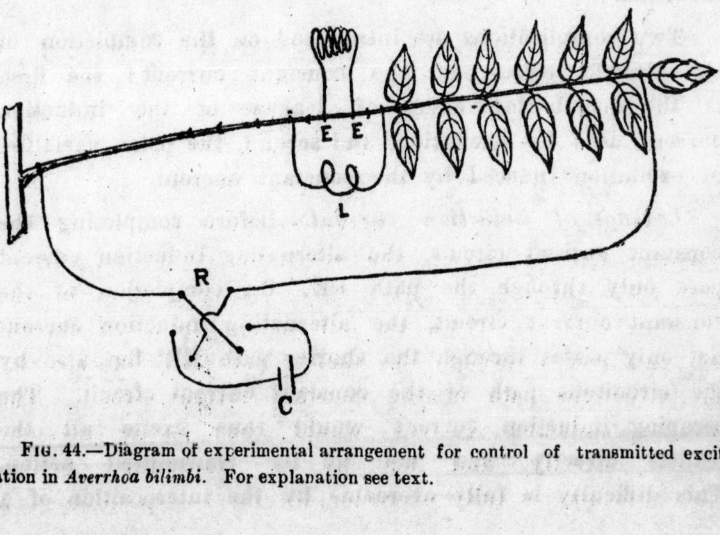

FIG. 44.—Diagram of experimental arrangement for control of transmitted excitation in *Averrhoa bilimbi*. For explanation see text.

for the purpose of this experiment and therefore removed. The intensity of the induction shock may be varied in the usual manner by removing the secondary coil nearer or farther from the primary. The duration of the shock is always maintained constant. On application of electric stimulus excitation is transmitted along the petiole, the distance of transmission depending on the intensity of stimulus. With feeble stimulus two pairs of leaflets may undergo an excitatory fall; with stronger stimulus the transmission is extended to the end of the petiole, and all the leaflets exhibit movements of closure. We shall now study the modifying influence of a constant current on conduction of excitation. C is an electric cell, R the reversing key by which the electric current could be sent from right to left or in the opposite direction. When the current is sent from right to the left, the excitatory impulse initiated at EE′ travels against the direction of the current in an 'up-hill' direction. When the current is reversed it flows in the petiole from left to right and the transmitted impulse travels with the current or in a 'down-hill' direction.

Two complications are introduced on the completion of the electric circuit of the constant current: the first, is the distributing effect of leakage of the induction current used for excitation, and second, the polar variation of excitation induced by the constant current.

Leakage of induction current.—Before completing the constant current circuit, the alternating induction current goes only through the path EE′. On completion of the constant current circuit, the alternating induction current not only passes through the shorter path EE′ but also by the circuitous path of the constant current circuit. The escaping induction current would thus excite all the leaflets directly and not by its transmitted action. This difficulty is fully overcome by the interposition of a choking coil which will be described below. A simpler, though less perfect, device may be employed to reduce and practically eliminate the leakage. This consists of a loop, L, of silver wire placed outside EE′. The leakage of induction current is thus diverted along this path of negligible resistance in preference to the longer circuit through the entire petiole, which has a resistance of several million ohms.

Polar action of current on excitability.—It is well known that an electric current induces a local depression of excitability at the point of entrance to the tissue, or at the anode, and an enhancement of excitability at the point of exit, or at the cathode. But the excitability is unaffected at a point equally distant from anode and cathode. This is known as the indifferent point. The exciting electrodes EE′ are placed at the indifferent point. But when the current enters on the right side, the terminal leaflets to the right have their excitability depressed by the proximity of anode, but the leaflets near the electrodes EE′, being at a distance from the anode are not affected by it. Moreover it will be shown that the enhanced conductivity conferred by the directive action of the current overpowers any depression of excitability in the terminal leaflets due to the proximity of the anode. I shall, for convenience, designate the transmission as 'up-hill', when excitation is propagated against the direction of the constant electric current, and 'down-hill' when transmitted with the direction of the current.

Transmission of excitation 'Up-hill': Experiment 38.—I shall give here an account of an experiment which may be taken as typical. I took a vigorous specimen of *Averrhoa bilimbi*, and applied a stimulus whose intensity was so adjusted that the propagated impulse brought about a fall of only two pairs of leaflets. This gave a measure of normal conduction without the passage of the current. The constant electric current was now sent from right to left. A necessary precaution is to increase the current gradually by means of a suitable potentiometer slide, to its full value. The reason for this will be given later. The intensity of the constant current employed was 1·4 micro-amperes. Now on exciting the petiole by the previous stimulus, the conducting power was found to be greatly enhanced. The excitatory impulse now reached the end of the petiole, and caused six pairs of leaflets to fall.

Transmission of excitation 'Down-hill': Experiment 39.—In continuation of the previous experiment, the constant electric current was reversed, its directions being now from left to right. Transmission of excitation was now in a down-hill direction. On applying the induction shock stimulus of the same intensity as before, the conducting power of the petiole was found to be abolished, none of the leaflets exhibiting any sign of excitation. This modification of the conducting power persists during the passage of the constant current. On cessation of the current the original conducting power is found to be restored. It will thus be seen that the power of conduction is capable of modification, and that the passage of an electric current of moderate intensity induces enhanced power of conduction in an 'up-hill' and diminished conductivity in a 'down-hill' direction.

ELECTRIC CONTROL OF NERVOUS IMPULSE IN ANIMALS.

In my 'Researches on Irritability of Plants' I have shown how intimately connected are the various physiological reactions in the plant and in the animal, and I ventured to predict that the recognition of this unity of response in plant and animal will lead to further discoveries in physiology in general. This surmise has been fully justified, as will be seen in the following experiments carried out on the nerve-and-muscle preparation of a frog. It is best to carry out the experiments with vigorous specimens; this ensures success, even in long continued experiments, which can then be repeated with unfailing certainty for hours. It is also an advantage to use a large frog for its relatively great length of the nerve.

Directive action of current on conduction of excitation in a nerve-and-muscle preparation: Experiment 40.—A preparation was made with a length of the spine and two nerves leading to the muscles. The specimen is supported in a suitable manner, and electric connections made with the toes, one for the entrance and the other for exit of the constant current. The current thus entered, say, by the left toe ascended the muscle and went up the nerve on the left side, and descended through the other nerve on the right side along the muscle and thence to the right toe. Before the passage of the constant electric current the spinal nerve was stimulated by an induction shock of definite intensity. The nervous impulse was conducted by the two nerves, one to the left and the other to the right, and caused a feeble twitch of the respective muscles. A feeble current of 1·5 micro-ampere was sent along the nerve-and-muscle circuit, ascending by the left and descending by the right side. It will be seen that excitation initiated at the spine is propagated 'against' the electric current on the left side, and 'with' the current on the right side. On repetition of previous electric stimulus the effect of directive action of current was at once manifested by the left limb being thrown into a state of strong tetanic contraction, whereas the right limb remained quiescent. By changing the direction of the constant current the induced enhancement of conductivity of the nerve was quickly transferred from the left to the right side, the depression or arrest of conduction being simultaneously transferred to the left side. Turning the reversing key one way or the other brought about *supra* or *non*-conducting state of the nerve, and this condition was maintained throughout the duration of the current.

I shall next describe a more perfect method for obtaining quantitative results both with plant and animal. In order to demonstrate the universality of the phenomenon, I next used *Mimosa pudica* instead of *Averrhoa*, for experiments on plants.

For determination of normal velocity of transmission of excitation and the induced variation of that velocity, I employed the automatic method of recording the velocity of transmission of excitation in *Mimosa*, where the excitatory fall of the motile leaf gave a signal for the arrival of the excitation initiated at a distant point. In this method the responding leaf is attached to a light lever, the writer being placed at right angles to it. The record is taken on a smoked glass plate, which during its descent makes an instantaneous electric contact, in consequence of which a stimulating shock is applied at a given point of the petiole. A mark in the recording plate indicates the moment of application of stimulus. After a definite interval the excitation is conducted to the responding pulvinus, when the excitatory fall of the leaf pulls the writer suddenly to the left. From the curve traced in this manner the time-interval between the application of stimulus and the initiation of response can be found, and the normal rate of transmission of excitation through a given length of the conducting tissue deduced. The experiment is then repeated with an electric current flowing along the petiole with or against the direction of transmission of excitation. The records thus obtained enable us to determine the influence of the direction of the current on the rate of transmission. I shall presently describe the various difficulties which have to be overcome before the method just indicated can be rendered practical.

The scope of investigation will be best described according to the following plan*:—

PART I.—INFLUENCE OF DIRECTION OF ELECTRIC CURRENT ON CONDUCTION OF EXCITATION IN PLANTS.

General method of experiment.

Effect of feeble current on velocity of transmission of excitation 'up-hill' or 'down-hill.'

Determination of variation of conductivity by the method of minimal stimulus and response.

The after-effect of current.

PART II.—INFLUENCE OF DIRECTION OF ELECTRIC CURRENT ON CONDUCTION OF EXCITATION IN ANIMAL NERVE.

The method of experiment.

Variation of velocity of transmission under the action of current.

Variation in the intensity of transmitted excitation.

PART I.—INFLUENCE OF DIRECTION OF CURRENT ON TRANSMISSION OF EXCITATION IN PLANT.

THE METHOD OF EXPERIMENT.

I may here say a few words of the manner in which the period of transmission can be found from the record given by my Resonant Recorder, fully described in my previous paper. The writer attached to the recording lever of this instrument is maintained by electromagnetic means in a state of to-and-fro vibration. The record thus consists of a series of dots made by the tapping writer, which is tuned to vibrate at a definite rate, say, 10 times per second. In a particular case whose record is given in Curve 1 (Fig. 46), indirect stimulus of electric stock was applied at a distance of 15 mm. from the responding

* For fuller account *see* Bose—'The influence of Homodromous and Heterodromous Electric Current on Transmission of Excitation in Plant and Animal.' Proc. R. S. B., Vol. 88, 1915.

pulvinus. There are 15 intervening dots between the moment of application of stimulus and the beginning of response; the time-interval is therefore 1·5 seconds. The latent period of the motile pulvinus is obtained from a record of direct stimulation; the average value of this in summer is 0·1 second. Hence the true period of transmission is 1·4 seconds for a distance of 15 mm. The velocity determined in this particular case is therefore 10·7 mm. per second.

Precaution has to be taken against another source of disturbance, namely, the excitation caused by the sudden commencement or the cessation of the constant current. I have shown elsewhere* that the sudden initiation or cessation of the current induces an excitatory reaction in the plant-tissue similar to that in the animal tissue. This difficulty is removed by the introduction of a sliding potentiometer, which allows the applied electromotive force to be gradually increased from zero to the maximum or decreased from the maximum to zero.

The experimental arrangement is diagrammatically shown in Fig. 45. After attaching the petiole to the recording lever, indirect stimulus is applied, generally speaking, at a distance of 15 mm. from the responding pulvinus. Stimulus of electric shock is applied in the usual manner, by means of a sliding induction coil. The intensity of the induction shock is adjusted by gradually changing the distance between the secondary and the primary, till a minimally effective stimulus is found. In the study of the effect of direction of constant current on conductivity, non-polarisable electrodes make suitable electric connections, one with the stem and the other with the tip of a sub-petiole, at a distance from each other of about 95 mm. The point of stimulation and the responding pulvinus are thus situated at a considerable distance from the anode or the cathode, in the indifferent region in which there is no

* BOSE—'Plant Response' (1906); 'Irritability of Plants' (1913).

polar variation of excitability. By means of a Pohl's commutator or reverser, the constant current can be maintained either "with" or "against" the direction of transmission of excitation. The transmission in the former case is

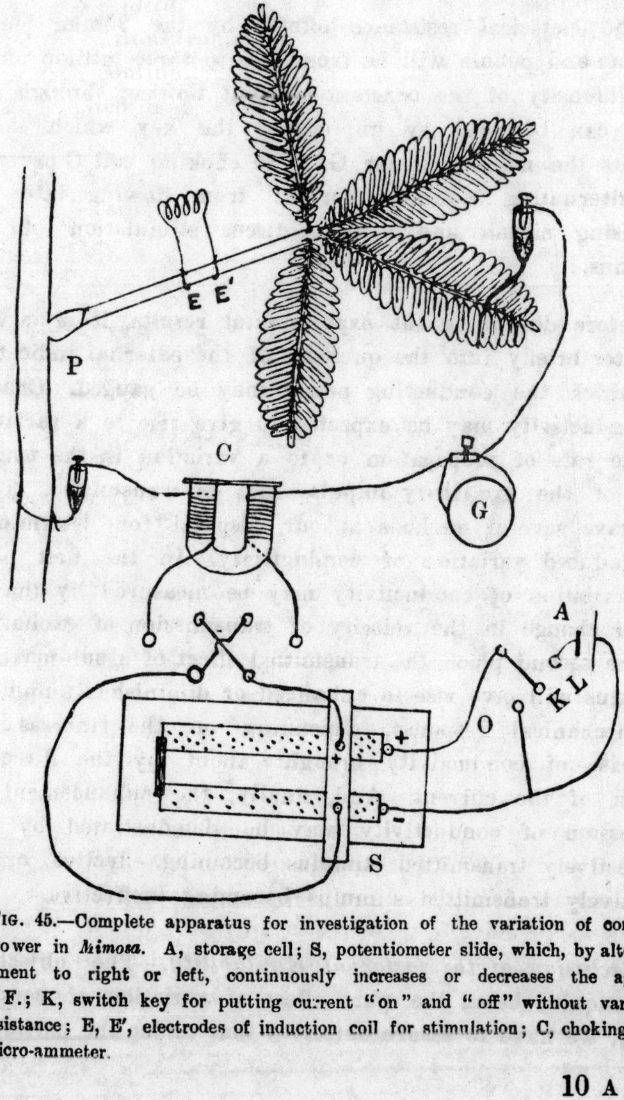

FIG. 45.—Complete apparatus for investigation of the variation of conducting power in *Mimosa.* A, storage cell; S, potentiometer slide, which, by alternate movement to right or left, continuously increases or decreases the applied E. M. F.; K, switch key for putting current "on" and "off" without variation of resistance; E, E', electrodes of induction coil for stimulation; C, choking coil; G, micro-ammeter.

"down-hill," and in the latter case "up-hill." Electrical connections are so arranged that when the commutator is tilted to the right, the transmission is down-hill, when tilted to the left, up-hill.

The electrical resistance offered by the 95mm. length of stem and petiole will be from two to three million ohms. The intensity of the constant current flowing through the plant can be read by unplugging the key which short-circuits the micro-ammeter G. The choking coil C prevents the alternating induction current from flowing into the polarising circuit and causing direct stimulation of the pulvinus.

Before describing the experimental results, it is as well to enter briefly into the question of the external indication by which the conducting power may be gauged. Change of conductivity may be expected to give rise to a variation in the rate of propagation or to a variation in the magnitude of the excitatory impulse that is transmitted. Thus we have several methods at our disposal for determining the induced variation of conductivity. In the first place the variation of conductivity may be measured by the induced change in the velocity of transmission of excitation. In the second place, the transmitted effect of a sub-maximal stimulus will give rise to enhanced or diminished amplitude of mechanical response, depending on the increase or decrease of conductivity brought about by the directive action of the current. And, finally, the enhancement or depression of conductivity may be demonstrated by the ineffectively transmitted stimulus becoming effective, or the effectively transmitted stimulus becoming ineffective.

Exclusion of the factor of Excitability.—The object of the enquiry being the pure effect of variation of conductivity, we have to assure ourselves that under the particular conditions of the experiment the complicating factor of polar variation of excitability is eliminated. It is to be remembered that excitatory transmission in *Mimosa* takes place by means of a certain conducting strand of tissue which runs through the stem and the petiole. In the experiment to be described, the constant current enters by the tip of the petiole and leaves by the stem, or *vice versâ*, the length of the intrapolar region being 95 mm. The point of application of stimulus on the petiole is 40 mm. from the electrode at the tip of the leaf. The responding pulvinus is also at the same distance from the electrode on the stem. The point of stimulation and region of response are thus at the relatively great distance of 40 mm. from either the anode or the cathode, and may therefore be regarded as situated in the indifferent region. This is found to be verified in actual experiments.

EFFECTS OF DIRECTION OF CURRENT ON VELOCITY OF TRANSMISSION.

A very convincing method of demonstrating the influence of electric current on conductivity consists in the determination of changes induced in the velocity of transmission by the directive action of the current. For this purpose we have to find out the true time required by the excitation to travel through a given length of the conducting tissue (1) in the absence of the current, (2) 'against' and (3) 'with' the direction of the current. The true time is obtained by substracting the latent period of the pulvinus from the observed interval between the stimulus and response. Now the latent period may not remain constant, but undergo change under the action of the polarising current. It has been shown that the excitability of the pulvinus does not undergo any change when it is situated in the middle or indifferent region. The following results show that under parallel conditions the latent period also remains unaffected :—

TABLE V.—SHOWING THE EFFECT OF ELECTRIC CURRENT ON THE LATENT PERIOD.

Specimens	I.	II.
	sec.	sec.
Latent period under normal condition	0·10	0·09
" " current from right to left	0·11	0·10
" " current from left to right	0·09	0·09

The results of experiments with two different specimens given above show that a current applied under the given conditions has practically no effect on the latent period, the slight variation being of the order of one-hundredth part of a second. This is quite negligible when the total period observed for transmission is, as in the following cases, equal to nearly 2 seconds.

Induced changes in the Velocity of Transmission.—Having found that the average value of the latent period in summer is 0·1 second, we next proceed to determine the influence of the direction of current on velocity.

Experiment 41.—As a rule, stimulus of induction shock was applied in this and in the following experiments on the petiole at a distance of 15 mm. from the responding pulvinus. The recording writer was tuned to 10 vibrations per second; the space between two succeeding dots, therefore, represents a time-interval of 0·1 second. The middle record, N in Fig. 46, is the normal. There are 17 spaces between the application of stimulus and the beginning of response. The total time is therefore 1·7 seconds, and by subtracting from it the latent period of 0·1 second we obtain the true time, 1·6 seconds. The normal velocity is found by dividing the distance 15 mm. by the true interval 1·6 seconds. Thus V = 15/1·6 = 9·4 mm. per second. We shall next consider the effect of current in modifying the normal velocity. The uppermost record (1) in Fig. 46 was taken under the action of an

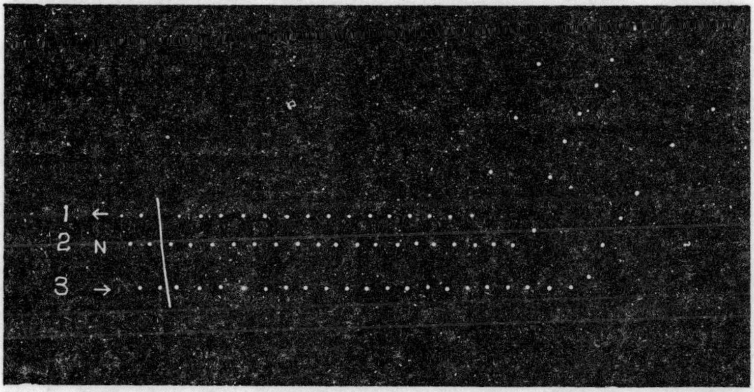

FIG. 46.—Record showing enhancement of velocity of transmission "up-hill" or against the current (uppermost curve) and retardation of velocity "down-hill" or with the current (lowest curve). N, normal record in the absence of current ← indicates "up-hill" and → "down-hill" transmission.

'up-hill,' or 'against' current of the intensity of 1·4 micro-ampères. It will be seen that the time interval is reduced from 1·7 seconds to 1·4 seconds; making allowance for the latent period, the velocity of transmission under 'up-hill' current $V_1 = 15/1 \cdot 3 = 11 \cdot 5$ mm. per second. In the lowest record (3) we note the effect of 'down-hill' current, the time-interval between stimulus and response being prolonged to 1·95 seconds and the velocity reduced to 8·1 mm. per second. The conclusion arrived at from this mechanical mode of investigation is thus identical with that derived from the electric method of conductivity balance referred to previously.

That is to say, *the passage of a feeble current modifies conductivity for excitation in a selective manner. Conductivity is enhanced* against, *and diminished* with, *the direction of the current.*

The minimum current which induces a perceptible change of conductivity varies somewhat in different specimens. The average value of this minimal current in autumn is 1·4 microampères. The effect of even a feebler current may be detected by employing a test stimulus which is barely effective.

TABLE VI.—SHOWING EFFECTS OF UP-HILL AND DOWN-HILL CURRENTS OF FEEBLE INTENSITY ON PERIOD OF TRANSMISSION THROUGH 15 MM.

Number.	Intensity of current in microampères.	Period for up-hill transmission.	Period for down-hill transmission.
1	1·4	14 tenths of a second	16 tenths of a second
2	1·4	13 " "	15 " "
3	1·6	19 " "	Arrest.
4	1·7	12 " "	14 tenths of a second.

Having demonstrated the effect of direction of current on the velocity of transmission, I shall next describe other methods by which induced variations of conductivity may be exhibited.

DETERMINATION OF VARIATION OF CONDUCTIVITY BY METHOD OF MINIMAL STIMULUS AND RESPONSE.

In this method we employ a minimal stimulus, the transmitted effect of which under normal conditions gives rise to a feeble response. If the passage of a current in a given direction enhances conductivity, then the intensity of transmitted excitation will also be enhanced; the minimal response will tend to become maximal. Or excitation which had hitherto been ineffectively transmitted will now become effectively transmitted. Conversely, depression of conductivity will result in a diminution or abolition of response. We may use a single break-shock of sufficient intensity as the test stimulus. It is, however, better to employ the additive effect of a definite number of feeble make-and-break shocks.

We may again employ additive effect of a definite number of induction shocks, the alternating elements of which are exactly equal and opposite. This is secured by causing rapid reversals of the primary current by means of a rotating commutator. The successive induction shocks of the secondary coil can thus be rendered exactly equal and opposite.

Experiment 42.—Working in this way, it is found that the transmitted excitation against the direction of current becomes effective or enhanced under 'up-hill' current. A current, flowing with the direction of transmission, on the other hand, diminishes the intensity of transmitted excitation or blocks it altogether.

Henceforth it would be convenient to distinguish currents in the two directions; the current in the direction of transmission will be distinguished as *Homodromous*, and against the direction of transmission as *Heterodromous*.

AFTER-EFFECTS OF HOMODROMOUS AND HETERODROMOUS CURRENTS.

The passage of a current through a conducting tissue in a given direction causes, as we have seen, an enhanced conductivity in an opposite direction. We may suppose this to be brought about by a particular molecular arrangement induced by the current, which assisted the propagation of the excitatory disturbance in a selected direction. On the cessation of this inducing force, there may be a rebound and a temporary reversal of previous molecular arrangement, with concomitant reversal of the conductivity variation. The immediate after-effect of a current flowing in a particular direction on conductivity is likely to be a transient change, the sign of which would be opposite to that of the direct effect. The after-effect of a heterodromous current may thus be a temporary depression, that of a homodromous current, a temporary enhancement of conductivity.

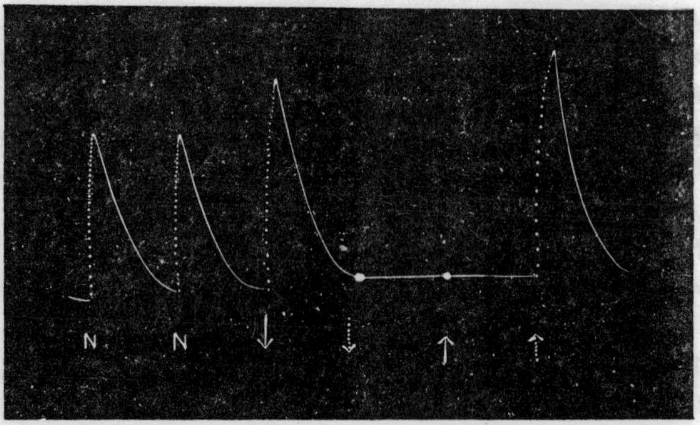

FIG. 47.—Direct and after-effect of heterodromous and homodromous currents. First two records, N, N, normal. ↓, enhanced transmission under heterodromous current; ⇣ arrest of conduction as an after-effect of heterodromous current. Next record ↑ shows arrest under homodromous current. Last record ⇡ shows enhancement of conduction greater than normal, as an after-effect of homodromous current. (Dotted arrow indicates the after-effect on cessation of a given current. ↑ homodromous and ↓ heterodromous current).

Experiment 43.—This inference will be found fully justified in the following experiment:—The first two responses are normal, after which the heterodromous current gave rise to an enhanced response. The depressing after-effect of a heterodromous current rendered the next response ineffective. The following record taken during the passage of the homodromous current exhibited an abolition of response due to induced depression of conductivity. Finally, the after-effect of the homodromous current is seen to be a response larger than the normal (Fig. 17). Those experiments show that the after-effect of cessation of a current in a given direction is a transient conductivity variation, of which the sign is opposite to that induced by the continuation of the current.

PART II—INFLUENCE OF DIRECTION OF ELECTRIC CURRENT ON CONDUCTION OF EXCITATION IN ANIMAL NERVE.

I shall now take up the question whether an electric current induced any selective variation of conductivity in the animal nerve, similar to that induced in the conducting tissue of the plant.

THE METHOD OF EXPERIMENT.

In the experiments which I am about to describe, arrangements were specially made so that (1) the excitation had not to traverse the polar region, and (2) the point of stimulation was at a relatively great distance from either pole. The fulfilment of the latter condition ensured the point of stimulation being placed at the neutral region.

In the choice of experimental specimens I was fortunate enough to secure frogs of unusually large size, locally known as "golden frogs" (*Rana tigrina*). A preparation was made of the spine, the attached nerve, the muscle and the tendon. The electrodes for constant current were applied at the extreme ends, on the spine and on the tendon (Fig. 48). The following are the measurements, in a typical case, of the different parts of the preparation.

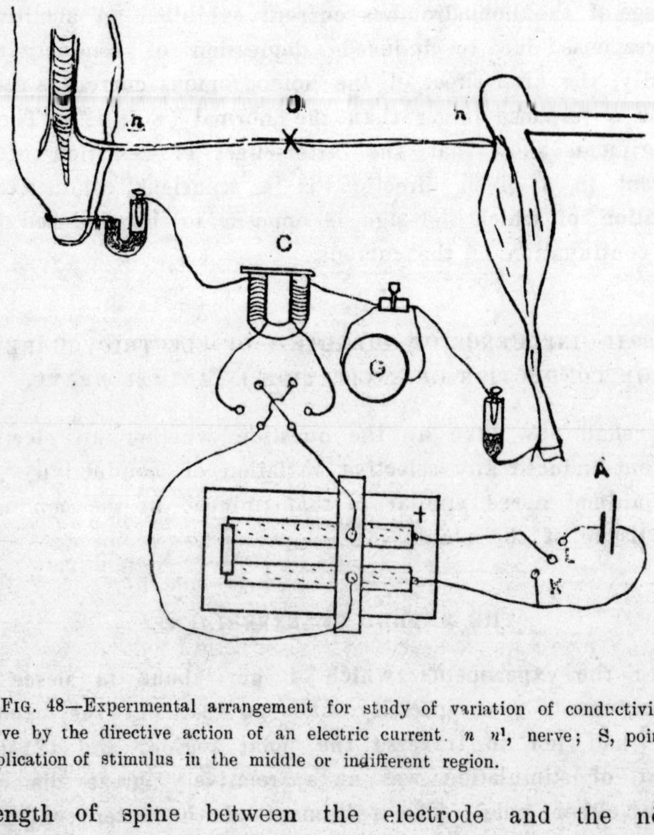

FIG. 48—Experimental arrangement for study of variation of conductivity of nerve by the directive action of an electric current. n n^1, nerve; S, point of application of stimulus in the middle or indifferent region.

Length of spine between the electrode and the nerve =40 mm.; length of nerve=90 mm.; length of muscle=50 mm.; length of tendon=30 mm. Stimulus is applied in all cases on the nerve, midway between the two electrodes this point being at a minimum distance of 100 mm. from either electrode. The point of stimulation is, therefore, situated at an indifferent region.

Great precautions have to be taken to guard against the leakage of current. The general arrangement for the experiment on animal nerve is similar to that employed for the corresponding investigations on the plant. The choking coil is used to prevent the stimulating induction current from getting round the circuit of constant current. The specimen is held on an ebonite support, and every part of the apparatus insulated with the utmost care.

VARIATION OF VELOCITY OF TRANSMISSION.

In the case of the conducting tissue of the plant a very striking proof of the influence of the direction of current on conductivity was afforded by the induced variation of velocity of transmission. Equally striking is the result which I have obtained with the nerve of the frog.

Experiment 44.—The experiments described below were carried out during the cold weather. The following records (Fig. 49), obtained by means of the pendulum myograph, exhibit the effect of the direction of current on

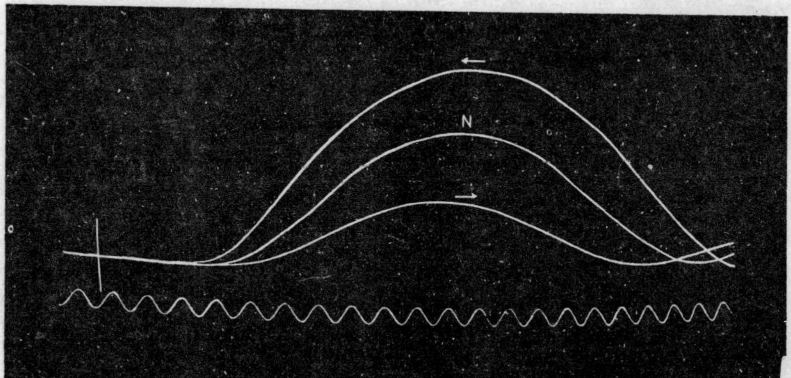

FIG. 49—Effect of heterodromous and homodromous current in inducing variation in velocity of transmission through nerve. N, normal record; upper record shows enhancement, and lower record retardation of velocity of transmission under heterodromous and homodromous currents, respectively.

the period of transmission through a given length of nerve. The latent period of muscle being constant, the variations in the records exhibit changed rates of conduction. The middle record is the normal, in the absence of any current. The upper record, denoted by the left-hand arrow, shows the action of a heterodromous current in shortening the period of transmission and thus enhancing the velocity above the normal rate. The lower record, denoted by the right-hand arrow, exhibits the effect of a homodromous current in retarding the velocity below the normal rate. I find that a very feeble heterodromous current is enough to induce a considerable increase of velocity, which soon reaches a limit. For inducing retardation of velocity, a relatively strong homodromous current is necessary. I give below a table showing the results of several experiments.

TABLE V.—EFFECT OF HETERODROMOUS AND HOMODROMOUS CURRENT OF FEEBLE INTENSITY ON VELOCITY OF TRANSMISSION.

Specimen.	Intensity of heterodromous current.	Acceleration above normal	Intensity of homodromous current.	Retardation below normal.
	microampère	per cent	microampères.	per cent.
1	0·35	16	1	20
2	0·7	13	1·5	19
3	0·8	18	2·0	14
4	0·8	11	2·0	13
5	1·0	18	2·5	12
6	1·5	15	3·0	40

VARIATION OF INTENSITY OF TRANSMITTED EXCITATION UNDER HETERODROMOUS AND HOMODROMOUS CURRENTS.

In the next method of investigation, the induced variation of intensity of transmitted excitation is inferred from the varying amplitude of response of the terminal muscle. Testing stimulus of sub-maximal intensity is applied at the middle of the nerve, where the constant current induces no variation of excitability. Stimulation is effected either by single break-shock or by the summated effects of a definite number of equi-alternating shocks, or by chemical stimulation

Experiment 45.—Under the action of feeble heterodromous current the transmitted excitation was always enhanced, whatever be the form of stimulation. This is seen illustrated in Fig. 50. Homodromous current on the other hand inhibited or blocked excitation (Fig. 51).

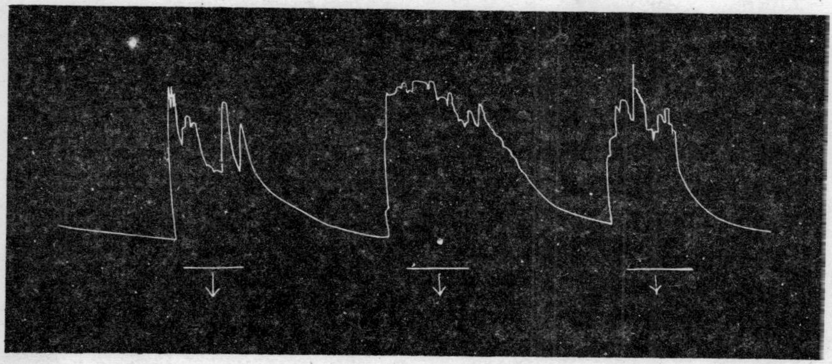

FIG. 50.—Ineffectively transmitted salt-tetanus becoming effective under heterodromous current, denoted by down-pointing arrow.

Complication due to variation of Excitabilitty of Muscle.—In experiments with the plant, there was the unusual advantage in having both the point of stimulation and the responding motile organ in the middle or indifferent region. Unfortunately this ideally perfect condition cannot be secured in experiments with the nerve-and-muscle preparation of the frog. It is true that the point of stimulation in this case is chosen to lie on the nerve at the middle or indifferent region. But the responding muscle is at one end, not very distant from the electrode applied on the tendon. It is, therefore, necessary to find out by separate experiments any variation of excitability that might be induced in the muscle by the proximity of either the anode or the cathode, and make allowance for such variation in interpreting the results obtained from investigations on variation of conductivity.

In the experimental arrangement employed, the hetrodromous current is obtained by making the electrode on the spine cathode and that on the tendon anode. The depressing influence of the anode in this case may be expected to lower, to a certain extent, the normal excitability of the responding muscle. Conversely, with homodromous current, the tendon is made the cathode and under its influence the muscle might have its excitability raised above the normal. These anticipations are fully supported by results of experiments. Sub-maximal stimulus of equi-alternating induction shock was directly applied to the muscle and records taken of (*1*) response under normal condition without any current, (*2*) response under heterodromous current, the tendon being the anode, and (*3*) response under homodromous current, the tendon being now made the cathode. It was thus found that under heterodromous current the excitability of the muscle was depressed, and under homodromous current the excitability was enhanced.

The effect of current on response to direct stimulation is thus opposite to that on response to transmitted excitation, as will be seen in the following Table.

TABLE VIII.—INFLUENCE OF DIRECTION OF CURRENT ON DIRECT AND TRANSMITTED EFFECTS OF STIMULATION.

Direction of current.	Transmitted excitation.	Direct stimulation.
Heterodromous current ...	Enhanced response ...	Depressed response
Homodromous current ...	Depressed response ...	Enhanced response.

The passage of a current, therefore, induces opposing effects on the conductivity of the nerve and the excitability of the muscle, the resulting response being due to their differential actions. Under heterodromous current a more intense excitation is transmitted along the nerve, on account of induced enhancement of conductivity. But this intense excitation finds the responding muscle in a state of depressed excitability. In spite of this the resulting response is enhanced (Fig. 50). The enhancement of conduction under heterodromous current is, in reality, much greater than is indicated in the record. Similarly, under homodromous current the depression of conduction in the nerve may be so great as to cause even an abolition of response, in spite of the enhanced excitability of the muscle (Fig. 51). The actual effects of current on conductivity are, thus, far in excess of what are indicated in the records.

AFTER-EFFECTS OF HETERODROMOUS AND HOMODROMOUS CURRENTS.

On the cessation of a current there is induced in the plant-tissue a transient conductivity change of opposite sign to that induced by the direct current (*cf. Expt. 43*). The same I find to be the case as regards the after-effect of current on conductivity change in animal nerve. Of this I only give a typical experiment of the direct and after-effect of homodromous current on salt-tetanus.

Experiment 46.—In this experiment sufficient length of time was allowed to elapse after the application of the salt

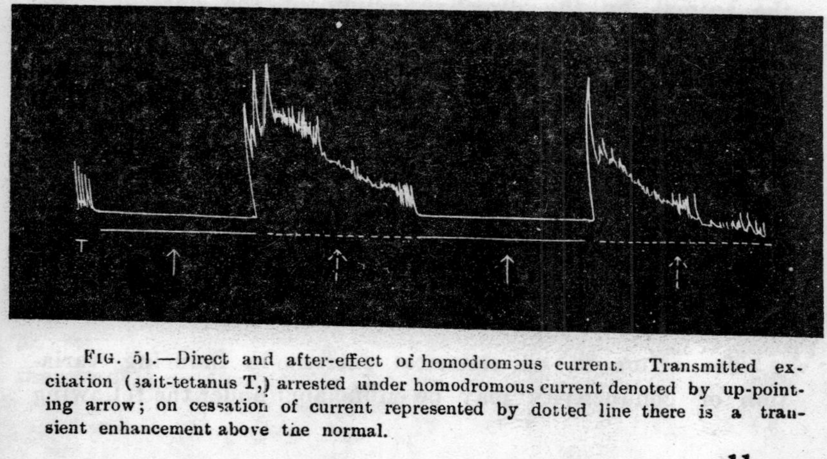

FIG. 51.—Direct and after-effect of homodromous current. Transmitted excitation (salt-tetanus T,) arrested under homodromous current denoted by up-pointing arrow; on cessation of current represented by dotted line there is a transient enhancement above the normal.

on the nerve, so that the muscle, in response to the transmitted excitation, exhibited an incomplete tetanus T. The homodromous current was next applied, with the result of inducing a complete block of conduction, with the concomitant disappearance of tetanus. The homodromous current was gradually reduced to zero by the appropriate movement of the potentiometer slide. The after-effect of homodromous current is now seen in the transient enhancement of transmitted excitation, which lasted for nearly 40 seconds. After this the normal conductivity was restored. Repetition of the experiment gave similar results (Fig. 51).

The results that have been given are only typical of a very large number, which invariably supported the characteristic phenomena that have been described.

It will thus be seen that with feeble or moderate current, conductivity is enhanced against the direction of the current and depressed or blocked with the direction of the current. Under strong current the normal effect is liable to undergo a reversal.

It has thus been shown that a perfect parallelism exists in the conductivity variation induced in the plant and in the animal by the directive action of the current. No explanation could be regarded as satisfactory which is not applicable to both cases. Now with the plant we are able to arrange the experimental condition in such a way that the factor of variation of excitability is completely eliminated. The various effects described about the plant-tissue are, therefore, due entirely to variation of conductivity. The parallel phenomena observed in the case of transmission of excitation in the animal nerve must, therefore, be due to the induced change of conductivity.

The action of an electrical current in inducing variation of conductivity may be enunciated under the following laws, which are equally applicable to the conducting issue of the plant and the nerve of the animal :—

LAWS OF VARIATION OF NERVOUS CONDUCTION UNDER THE ACTION OF ELECTRIC CURRENTS.

1. THE PASSAGE OF A CURRENT INDUCES A VARIATION OF CONDUCTIVITY, THE EFFECT DEPENDING ON THE DIRECTION AND INTENSITY OF CURRENT.
2. UNDER FEEBLE INTENSITY, HETERODROMOUS CURRENT ENHANCES, AND HOMODROMOUS CURRENT DEPRESSES, THE CONDUCTION OF EXCITATION.
3. THE AFTER-EFFECT OF A FEEBLE CURRENT IS A TRANSIENT CONDUCTIVITY VARIATION, THE SIGN OF WHICH IS OPPOSITE THAT INDUCED DURING THE CONTINUATION OF CURRENT.

SUMMARY.

The variation of conductivity induced by the directive action of current has been investigated by two different methods :—

(1) The method in which the normal speed and its induced variation are automatically recorded ;

(2) That in which the variation in the intensity of transmitted excitations is gauged by the varying amplitudes of resulting responses.

The great difficulty arising from leakage of the exciting induction current into the polarising circuit was successfully overcome by the interposition of a choking coil.

The following summarises the effects of direction and intensity of an electric current, on transmission of excitation through the conducting tissue of the plant.

The velocity of transmission is enhanced against the direction of a feeble current, and retarded in the direction of the current.

Feeble heterodromous current enhances conductivity, homodromous current, on the other hand, depresses it.

Ineffectively transmitted excitation becomes effectively transmitted under heterodromous current. Effectively transmitted excitation, on the other hand, becomes ineffectively transmitted under the action of homodromous current.

The after-effect of a current is a transient conductivity change, the sign of which is opposite to that induced during the passage of current. The after-effect of a heterodromous current is, thus, a transient depression, that of homodromous current, a transient enhancement of conductivity.

The characteristic variations of conductivity induced in animal nerve by the direction and intensity of current are in every way similar to those induced in the conducting tissue of the plant.

These various effects are demonstrated by the employment of not one, but various kinds of testing stimulus, such as the excitation caused (1) by a single break-induction shock or (2) by a series of equi-alternating tetanising shocks or (3) by chemical stimulation.

VIII.—EFFECT OF INDIRECT STIMULUS ON PULVINATED ORGANS

By

SIR J. C. BOSE,

Assisted by

GURUPRASANNA DAS, L.M.S.

THE leaf of *Mimosa pudica* undergoes an almost instantaneous fall when the stimulus is applied directly on the pulvinus which is the responding organ. The latent period, *i.e.*, the interval between the application of stimulus and the resulting response is about 0·1 second. Indirect stimulus, *i.e.*, application of stimulus at a distance from the pulvinus, also causes a fall of the leaf; but a longer interval will elapse between the incidence of stimulus and the response; for it will take a definite time for the excitation to be conducted through the intervening tissue. I have already shown that this conduction of excitation in plant is analogous to the transmission of nervous impulse in animal.

The power of conduction varies widely in different plants. In the petiole of *Mimosa pudica* the velocity may be as high as 30 mm. per second. In the stem the velocity is considerably less, *i.e.*, about 6 mm. per second in the longitudinal direction; but conduction across the stem is a very much slower process. In the petiole of

Averrhoa the longitudinal velocity is of the order of 1 mm. per second.

DUAL CHARACTER OF THE TRANSMITTED IMPULSE.

The record of the transmitted effect of stimulus is found to exhibit a remarkable preliminary variation. This was detected by my delicate recorders, which gave magnifications from fifty to hundred times. I shall give a detailed account of a typical experiment carried out with *Averrhoa carambola*, which will bring out clearly the characteristic effects of Indirect Stimulus.

Experiment 47.—Stimulus of electric shock applied at a point on the long petiole of *Averrhoa* causes successive fall of pairs of leaflets. In the experiment to be described one

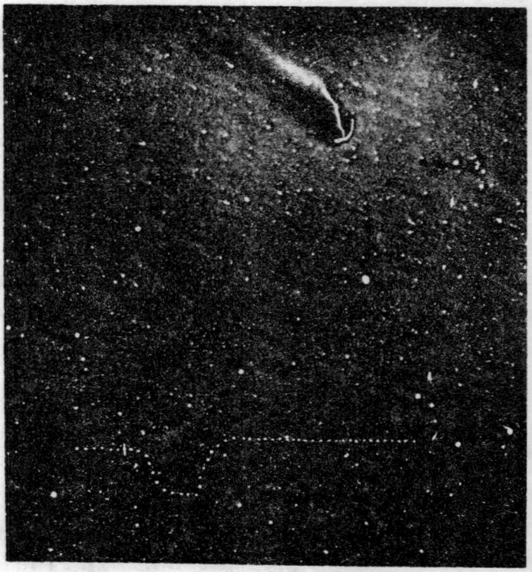

FIG. 52.—Effect of indirect Stimulus on leaflet of *Averrhoa carambola*. Stimulus was applied at the short vertical line. Successive dots at intervals of one second. Note the *positive* response preceding the *negative*.

of the leaflets of the plant was attached to the recorder. Stimulus was applied at a distance of 50 mm. The successive dots in the record are at intervals of a second. It will be noticed that two distinct impulses—a *positive* and a *negative*—were generated by the action of Indirect Stimulus. The positive impulse reached the responding organ after 1·5 second and caused an erectile movement. The velocity of the positive impulse in the present case is 33 mm. per second. The normal excitatory negative impulse reached the motile organ 44 seconds after the application of stimulus, and caused a very rapid fall of the leaflet, the fall being far more pronounced than the positive movement of erection (Fig. 52). In this and in all subsequent records, the positive and negative responses offer a great contrast. The movement in response to positive reaction is slow, whereas that due to negative reaction is very abrupt, almost 'explosive,' the successive dots being now very wide apart. As regards the velocity of impulse the relation is reversed, the positive being the quicker of the two. In the present case, the velocity of the excitatory *negative* impulse is 1·1 mm. per second, as against 33 mm. of the *positive* impulse.

The negative impulse is due to the comparatively slow propagation of the excitatory protoplasmic change, which brings about a diminution of turgor in the pulvinus and fall of the responding leaflet. The erectile movement of the leaflet by the positive impulse must be due to an increase of turgor, brought on evidently, by the forcing in of water. This presupposes a forcing out of water somewhere else, probably at the point of application of stimulus. It may be supposed that an active contraction occurred in plant-cells under direct stimulus, in consequence of which water was forced out giving rise to a hydraulic wave. On this supposition the positive impulse is to be regarded as hydro-mechanical. I have, however, not yet been able to devise a direct experimental test to settle the question.

EFFECT OF DISTANCE OF APPLICATION OF STIMULUS.

In the last experiment the stimulus was applied at the moderate distance of 50 mm. Let us now consider the respective effects, first, of an increase, and second, of a decrease of the intervening distance. In a tissue whose conducting power is not great, the excitatory impulse is weakened, even to extinction in transmission through a long distance. Thus the negative impulse may fail to reach the responding organ, when the stimulus is feeble or the intervening distance long or semi-conducing. Hence, under the above conditions, stimulus applied at a distance will give rise only to a positive response.

A reduction of the intervening distance will give rise to a different result. As the negative response is the more intense of the two, the feeble positive will be masked by the superposed negative. The separate exhibition of the two responses is only possible by a sufficient lag of the negative impulse behind the positive. This lag increases with increase of length of transmission and decreases with the diminution of the length. Hence the application of stimulus near the responding organ will give rise only to a negative response, in spite of the presence of the positive, which becomes masked by the predominant negative.*

These inferences have been fully borne out by results of experiments carried out with various specimens of plants under the action of diverse forms of stimuli. In all cases, application of stimulus at a distance causes a pure positive response; moderate reduction of the distance induces a diphasic response—a positive followed by a negative; further

* *Cf.* BOSE—"Plant Response," p. 535; "Comparative Electro-Physiology," p. 34; "Irritability of Plants," p. 196.

diminution of distance gives rise to a resultant negative response, the positive being masked by the predominant negative.

From what has been said it will be understood that the exhibition of positive response is favoured by the conditions, that the transmitting tissue should be semi-conducting, and the stimulus feeble. It is thus easier to exhibit the positive effect with the feebly conducting petiole of *Averrhoa* than with the better conducting petiole of ***Mimosa.*** It is, however, possible to obtain positive response in the ***Mimosa*** by application of indirect stimulus to the stem in which conduction is less rapid than in the petioles.

TABLE IX.—PERIODS OF TRANSMISSION OF POSITIVE AND NEGATIVE IMPULSES IN THE PETIOLE OF *AVERRHOA* AND STEM OF *MIMOSA*.

No.	Specimen.	Distance in mm.	Stimulus.	Transmission period for positive impulse.	Transmission period for negative impulse.
1	*Averrhoa* ...	70	Thermal ...	22 secs.	65 secs.
2	" ...	130	" ..	40 "	95 "
3	" ...	10	Induction-shock	6 "	20 "
4	" ...	20	" ...	14 "	48 "
5	" ...	35	Chemical ...	21 "	50 "
6	*Mimosa* ...	5	Induction-shock	0·5 "	12 "
7	" ...	10	" ...	0·6 "	9·4 "
8	" ...	20	" ...	1·1 "	10 "
9	" ...	60	" ...	2 "	29 "
10	" ...	35	Chemical ...	5 "	17 "

EFFECTS OF DIRECT AND INDIRECT STIMULUS.

From the results given in course of the Paper we are able to formulate the following laws about the effects of Direct and Indirect Stimulus on pulvinated organs :—

1. EFFECT OF ALL FORMS OF DIRECT STIMULUS IS A DIMINUTION OF TURGOR, A CONTRACTION AND A NEGATIVE MECHANICAL RESPONSE.
2. EFFECT OF INDIRECT STIMULUS IS AN INCREASE OF TURGOR, AN EXPANSION AND A POSITIVE MECHANICAL RESPONSE.

3. PROLONGED APPLICATION OF INDIRECT STIMULUS OF MODERATE INTENSITY GIVES RISE TO A DIPHASIC, POSITIVE MECHANICAL RESPONSE FOLLOWED BY THE NEGATIVE.

4. IF THE INTERVENING TISSUE BE HIGHLY CONDUCTING, THE TRANSMITTED POSITIVE EFFECT BECOMES MASKED BY THE PREDOMINANT NEGATIVE.

The laws of Effects of Direct and Indirect stimulus hold good not merely in the case of sensitive plants, but universally for all plants. This aspect of the subject will be treated in fuller detail in later Papers of this series.

IX.—MODIFYING INFLUENCE OF TONIC CONDITION ON RESPONSE

By

SIR J. C. BOSE

Assisted by

GURUPRASANNA DAS.

IN experiments with different pulvinated organs, great difference is noticed as regards their excitability. If electric shock of increasing intensity from a secondary coil be passed through the pulvini of *Mimosa*, *Neptunia*, and *Erythrina* arranged in series, it would be found that *Mimosa* would be the first to respond; a nearer approach of the secondary coil to the primary would be necessary for *Neptunia* to show sign of excitation. *Erythrina* would require a far greater intensity of electric shock to induce excitatory movement. Organs of different plants may thus be arranged, according to their excitability, in a vertical series, the one at the top being the most excitable. The specific excitability of a given organ is different in different species.

In addition to this characteristic difference, an identical organ may, on account of favourable or unfavourable conditions, exhibit wide variation in excitability. Thus under favourable conditions of light, warmth and other factors, the excitability of an organ is greatly enhanced. In the absence of these favourable tonic conditions the excitability is depressed or even abolished. I shall, for convenience, distinguish the different tonic conditions of the plant as *normal*, *hyper-tonic* and *sub-tonic*. In the first case, stimulus of moderate intensity will induce excitation ; in the second, the excitability being exceptionally high, very feeble stimulus will be found to precipitate excitatory reaction. But a tissue in a *sub-tonic* condition will require a very strong stimulus to bring about excitation. The excitability of an organ is thus determined by two factors : the specific excitability, and the tonic condition of the tissue.

THEORY OF ASSIMILATION AND DISSIMILATION.

A muscle contracts under stimulus ; this is assumed to be due to some explosive chemical change which leaves the tissue in a condition less capable of functioning, or in a condition below par. Herring designates this as a process of *dissimilation*. The excitability of the muscle is restored after suitable periods of rest, by the opposite metabolic change of *assimilation*. "Assimilation and Dissimilation must be conceived as two closely interwoven processes, which constitute the metabolism (unknown to us in its intrinsic nature) of the living substance. Excitability diminishes in proportion with the duration of D-stimulus, or, as it is usually expressed, the substance *fatigues* itself. It is perfectly intelligible that a progressive fatigue and decrement of the magnitude of contraction must ensue. The only point that is difficult to elucidate is the initial staircase increment of the twitches, more especially in excised, bloodless muscle, which seems in direct contradiction with the previous theory."*

With reference to Herring's theory given above, Bayliss in his "Principles of General Physiology" (1915), page 377

* BIEDERMANN—Electro-Physiology (English Translation), Vol 1, pp. 83, 84, 5; Macmillan & Co.

says, "In the phenomenon of metabolism, two processes must be distinguished, the building up of a complex system or substance of high potential energy, 'anabolism,' and the breaking down of such a system 'catabolism,' giving off energy in other forms. The tendency of much recent work, however, is to throw doubt on the universality of this opposition of anabolism and catabolism as explanatory of physiological activity in general."

The results obtained with the response of plants to stimulus may perhaps throw some light on the obscurities that surround the subject. They show that the two processes may be present simultaneously, and that the 'down' change induced by stimulus may, in certain instances, be more than compensated by the 'up' change.* I shall, for convenience, designate the physico-chemical modification, associated with the excitatory negative mechanical and electrical response of plants, as the "D" change; this is attended by run down of energy. The positive mechanical and electrical response must therefore connote opposite physico-chemical change, with increase of potential energy. This I shall designate as the "A" change, which by increasing the latent energy, enhances the functional activity of the tissue. That stimulus may give rise simultaneously to both A, and D, effects, finds strong support in the dual reactions exhibited in plant-response. Under indirect stimulus, the two responses are seen separately, the more intense negative following the feeble positive. When by the reduction of the intervening distance, stimulus is made direct, the resultant response, as previously stated, is negative; and this is due not to the total absence of the positive but to its being masked by the predominant negative. Let us next

* In the response of inorganic matter I have obtained records of positive, diphasic and negative responses. It would perhaps be advisable to refer the 'A' and 'D' effects, to physico-chemical change. The simultaneous double reaction, combination and decomposition, is of frequent occurrence in many chemical changes.

consider the question of unmasking this positive element in the resultant negative response.

UNMASKING OF THE POSITIVE EFFECT.

Under favourable conditions of the environment, the excitability of the organs is at its maximum. A given stimulus will bring about an intense excitation, and the 'down' D-change will therefore be very much greater than the A-change. Let us now consider the case at the opposite extreme where, owing to unfavourable condition, the excitability is at its lowest. Under stimulus the excitatory D-change will now be relatively feeble compared to the A-change, by which the potential energy of the system becomes increased. In such a case successive stimuli will increase the functional activity of the tissue, and bring about staircase response. Biedermann mentions the staircase response of *excised bloodless muscle* as offering difficulty of explanation. It is obvious that the physiological condition of the excised muscle must have fallen below par. The staircase response in such a tissue is thus explained from considerations that have just been adduced.

The results obtained with *Mimosa* not only corroborate them, but add incontestable proof of the simultaneous existence of both A and D changes. The physiological condition of a plant, *Mimosa* for example, is greatly modified by the favourable or unfavourable condition of the environment. In a hyper-tonic condition its excitability becomes very great; in this condition the plant responds to its maximum even under very feeble stimulus. Here the D-change is relatively great, and successive responses are apt to show sign of fatigue.

But the plant in a sub-tonic condition will exhibit feeble or no excitation. The D-change will be absent while the A-change will take place under the action of stimulus. This, by increasing the potential energy, will enhance the functional activity of the tissue.

Staircase response in Mimosa: Experiment 48.—The theoretical considerations will be found experimentally verified in the record obtained with a specimen of *Mimosa* in a sub-tonic condition (Fig. 53). Owing to the lack of favourable 'tone' the leaf was relaxing as seen in the first part of the curve. The stimulus of electric shock, applied at the thick dot in the curve slanting downwards, gave no response but raised the tone of the tissue by arresting the growing relaxation. Subsequent stimuli gave rise to staircase responses. Stimulus has, through the A-effect, raised the functional activity of the tissue to a maximum.

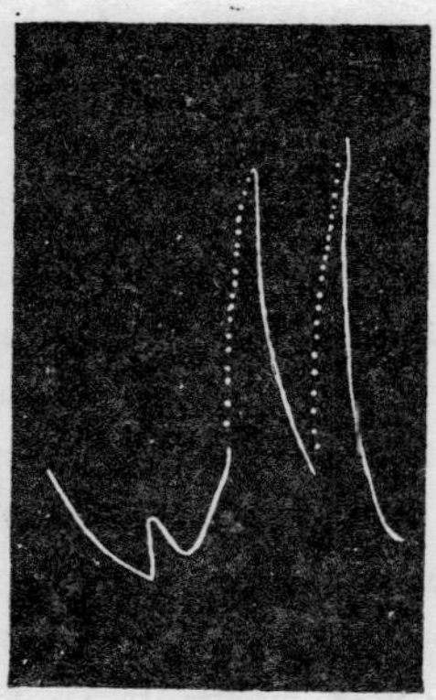

FIG. 53.—Record showing the effect of stimulus modifying tonicity and producing staircase effect. (*Mimosa.*)

ARTIFICIAL DEPRESSION OF TONIC CONDITION AND MODIFICATION OF RESPONSE.

It has been shown that while favourable tonic condition has the effect of raising the excitability and enhancing the negative response with the associated D-change, a condition of sub-tonicity, on the other hand, induces depression of excitability, a diminution of negative response and of the attendant D-change. In this condition the positive element in the response with the A-change will come into greater prominence. These considerations led me to experiment with specimens exhibiting increasing sub-tonicity, with a view of ummasking the positive element in the response, *i.e.*, the A-change. In the last experiment a specimen was found which happened to be in a sub-tonic condition on account of the unfavourable condition of its surroundings. I was next desirous of securing specimens in which I could induce increasing sub-tonicity at will.

I have shown (*Expt. 23*) that a detached branch of *Mimosa* can be kept alive for several days with the cut end immersed in water. In this condition the pulvinus retains its sensitiveness for more than two days. The excitability undergoes a continuous decline and is abolished about the fiftieth hour. Isolation from the parent organism thus causes a continuous depression of the tonic condition of the specimen. The case is somewhat analogous to the depression of excitability in an excised bloodless muscle. It is thus possible to secure specimens of varying degrees of sub-tonicity. A specimen that has been detached for six hours will exhibit a slight amount of depression, while a different specimen isolated for twenty-four hours will occupy a very much lower position in the scale of tonicity.

Experiment 49.—The staircase response of *Mimosa* given in figure 53 was obtained with the stimulus of induction shock. In order to establish a wider generalisation I now used the stimulus of light given by an arc lamp. There may be a difficulty on account of the diurnal movement of *Mimosa*; the leaf, generally speaking, has a movement in a downward direction from morning till noon, after which there is a comparative state of rest. It is better to choose the time of noon for experiment. In any case the response to stimulus is very abrupt and in strong contrast with the slow diurnal movement A horizontal pencil of light was thrown upwards by means of a small mirror and made to fall on the lower half of a pulvinus of the *Mimosa* leaf. The excitatory down movement is followed by recovery on the cessation of light. The intensity of stimulus can be modified by varying the intensity of light. I took for my first series of experiments a specimen that had been isolated for six hours. Stimulation was caused by successive applications of light for 25 seconds at intervals of

3 minutes. Figure 54 shows how the functional activity of the sub-tonic specimen is enhanced by stimulus, the successive responses thus exhibiting the staircase effect.

FIG. 54.

FIG. 55.

FIG. 54.—Staircase response in sub-tonic *Mimosa*.
FIG. 55.—Positive, diphasic and negative response under successive stimulation.

POSITIVE RESPONSE IN SUB-TONIC SPECIMEN.

Experiment 50.—A still lower degree of sub-tonicity was ensured by keeping the specimen in an isolated condition for 12 hours. Stimulus of light for 20 seconds' duration was applied at intervals of 2 minutes. In the record (Fig. 55) the first two responses, not shown, were purely positive. The third exhibited a positive A-effect, followed by the negative response D-effect. The A-effect is thus seen fully unmasked. In subsequent responses the A-effect became more and more overshadowed by the D-effect. At the third response the masking is complete and the excitatory negative response is at its maximum. The record of staircase effect (Fig. 54) also exhibits a preliminary positive twitch at the beginning of the series, which disappeared after the second response.

The modifying influence of tonic condition on response I find to be of universal occurrence. In vigorous specimens the electric response to stimulation is *negative;* but tissues in sub-tonic condition give *positive* response and after long-continued stimulation the abnormal positive is converted into the normal *negative.* It is very interesting that under condition of sub-tonicity diverse expressions of physiological reaction exhibit similar change of sign of normal response. Thus in my measurement of the velocity of transmission of excitation in the conducting tissue of *Mimosa*, I find that, when the tissue is in an optimum condition, exhibiting high velocity of transmission, excessive stimulus has the effect of diminishing the conducting power. But in a depressed condition of the tissue the effect is precisely the opposite. Thus in a given case the velocity of transmission was low ; strong electric stimulation enhanced the rate by 33 per cent. In extreme cases of sub-tonicity, where the conducting power was in abeyance, the excessive stimulus caused by wound not only restored the power of conduction but raised the velocity of transmission to 25 mm. per second (*Expt. 37*).

SUMMARY.

The excitability of a plant is found to be modified by its tonic condition.

A sub-tonic specimen of *Mimosa*, like an excised bloodless muscle, shows a preliminary staircase response. Stimulus induces simultaneously both "A" and "D" effects, with their attendant positive and negative reactions.

A tissue in optimum condition exhibits only the resultant negative response, the comparatively feeble positive being masked by the predominant negative. With decline of tone, the "D" effect diminishes and we get "A" effect unmasked.

In extreme sub-tonic specimen, we get first only the "A" effect, with its positive response. Successive stimulation converts the pure positive into diphasic and ultimately into normal negative response.

PART II.

GROWTH AND ITS RESPONSIVE VARIATIONS.

X.—THE HIGH MAGNIFICATION CRESCOGRAPH FOR RESEARCHES ON GROWTH*

By

SIR J. C. BOSE,

Assisted by

GURUPRASANNA DAS, L.M.S.

In discussing the difficulties connected with investigations relating to longitudinal growth and its variations, special stress must be laid on the importance of maintaining external conditions absolutely constant. This constancy can only be maintained in practice for a short time. Lengthy periods of observation, moreover, introduce the uncertainty of complication arising from spontaneous variation of growth. The possibility of accurate investigation, therefore lies in reducing the period of the experiment to a few minutes during which we have to determine the normal rate of growth and its variation under a given changed condition. This would necessitate the devising of a method of very high magnification for record of the rate of growth.*

With auxanometers now in use, which give a magnification of about twenty times, it takes nearly four hours to determine the influence of changed condition in inducing

* A short account of my researches with the High Magnification Crescograph has been published in the *Proceedings* of the Royal Society. I shall in the following Papers give a detailed account of my investigations on growth and on allied phenomena.

variation of growth. It will be seen that if we succeeded in enhancing magnification from twenty to ten thousand times, the necessary period for experiment would be reduced from four hours to thirty seconds. The importance of securing a magnification of this order is sufficiently obvious.

The problem of high magnification was first solved by my Optical Lever.* The tip of the growing organ was attached to the short arm of a lever, the axis of which carried a small mirror; in this way it was possible to obtain a magnification of a thousand times. The magnified movement of growth was followed with a pen on a revolving drum. The record laboured under the disadvantage of not being automatic. This defect was overcome by the use of the photographic method which however entailed the inconvenience and discomfort of a dark room.

I have, for the past six years, been working with a different method, which has now been brought to a great state of perfection. The problem to be solved was the devising of a direct method of high magnification and the automatic record of the magnified rate of growth.

METHOD OF HIGH MAGNIFICATION.

The magnification in my Crescograph is obtained by a compound system of two levers. The growing plant is attached to the short arm of a lever, the long arm of which is attached to the short arm of the second lever. If the magnification by the first lever be m, and that by the second, n, the resulting magnification would be mn.

The practical difficulties met with in carrying out this idea are very numerous. It will be understood that just as the imperceptible movement is highly magnified by the

* BOSE—"Plant Response," p. 412.

compound system of levers, the various errors and difficulties are likely to be magnified in the same proportion. The principal difficulties met with were due: (1) to the weight of the compound lever which exerted a great tension on the growing plant, (2) to the yielding of flexible connections by which the plant was attached to the first lever, and the first lever to the second, and (3) to the friction at the fulcrums.

Weight of the Lever.—As the first lever is to exert a pull on the second, it has to be made rigid. The second lever serves as an index, and can therefore be made of fine glass fibre. The securing of rigidity of the first lever entails large cross section and consequent weight, which exerts considerable tension on the plant. Excessive tension greatly modifies growth; even the weight of the index used in self-recording auxanometers is found to modify the normal rate of growth. The weight of the levers introduces an additional difficulty in the increased friction at the fulcrums, on account of which there is an obstruction of the free movement of the recording arm of the lever. The conditions essential for overcoming these difficulties therefore are: (1) construction of a very light lever possessing sufficient rigidity, and (2) arranging the levers in such a way that the tension on the plant may be reduced to any extent, or even eliminated.

I found in *navaldum*, an alloy of aluminium, a light material possessing sufficient rigidity. The first lever is constructed out of a thin narrow sheet 25 cm. in length; it has, as explained before, to be fairly rigid in order to exert a pull on the second without undergoing any bending; this rigidity is secured by giving the thin narrow plate of the lever a T-shape. The first lever balances, to a certain extent, the second. Finer adjustments are made by means of an adjustable counterpoise B, at the end of the levers. By this means the tension on the plant can be greatly reduced; or a constant tension may be exerted by means of a weight T (Fig. 56). In my later type

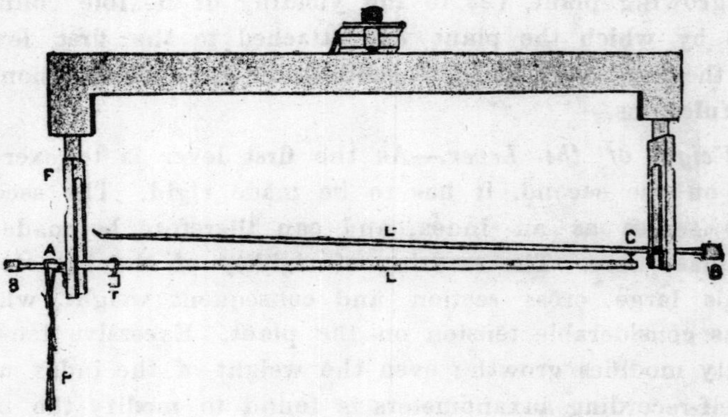

FIG. 56.—Compound lever. P, plant attached to short arm of lever L; T, weight exerting tension; C, connecting link; L,' second lever with bent tip for record; B, B, balancing counterpoise. Fork F, carries at its side two conical agate cups, on which lever rests by two pin-points. (From a photograph.)

of the apparatus the plant connection is made to the right, instead of the left side of the first fulcrum. This gives certain practical advantages. The second lever is then made practically to balance the first, only a very slight weight being necessary for exact counterpoise. The reduction of total weight thus secured reduces materially the friction at the fulcrum with great enhancement of efficiency of the apparatus.

The second or the recording lever has a normal excursion through 8 cm. on the recording surface, which is a very thin sheet of glass 8×8 cm. coated with a layer of smoke. As the recording lever is about 40 cm. in length, the curvature in the record is slight, and practically negligible in the middle portion of 4 cm. The dimensions given allow a magnification of ten thousand times. A far more compact apparatus is made with 15 cm. length of levers. This gives a magnification of a thousand times.

AUTOMATIC RECORD OF THE RATE OF GROWTH.

Another great difficulty in obtaining an accurate record of the curve of growth arises from the friction of contact of the bent tip of the writing lever against the recording surface. This I was able to overcome by an oscillating device by which the contact, instead of being continuous, was made intermittent. The smoked glass plate, G, is made to oscillate, to and fro, at regular intervals of time, say one second. The bent tip of the recording lever comes periodically in contact with the glass plate during its extreme forward oscillation. The record would thus consist of a series of dots, the distance between successive dots representing magnified growth during a second.

The drawback in connection with the obtaining of record on the oscillating plate lies in the fact that if the plate approaches the recording point with anything like suddenness, then the stroke on the flexible lever causes an after-oscillation ; the multiple dots, thus produced, spoil the record. In order to overcome this, a special contrivance is necessary, by which the speed of approach of the plate should be gradually reduced to zero at contact with the recording point. The rate of recession should, on the other hand, continuously increase from zero to maximum. The recording point will in this manner be gently pressed against the glass plate, marking the dot, and then gradually set free. It was only after strict observance of these conditions that the disturbing effect of after-vibration of the lever could be obviated

This particular contrivance consists of an eccentric rod actuated by a rotating wheel. A cylindrical rod is supported eccentrically, so that semi-rotation of the eccentric causing a pull on the crank K (Fig. 57) pushes the plate

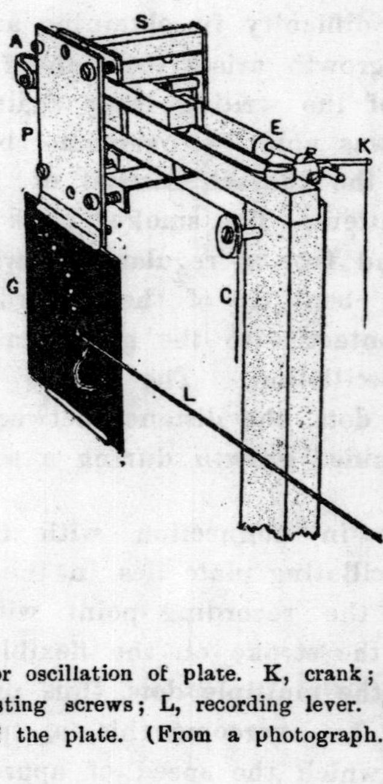

FIG. 57.—Eccentric for oscillation of plate. K, crank; S, slide; P, holder for glass plate G. A, adjusting screws; L, recording lever. Clock releases string C for lateral movement of the plate. (From a photograph.)

carrier gradually forward. On the return movement of the eccentric, a light antagonistic spring makes the plate recede. The rate of the movement of the crank itself is further regulated by the device of the revolving wheel. This is released periodically by clockwork at intervals of one, two, five, ten, or fifteen seconds respectively, according to the requirements of the experiment. The complete apparatus is shown in figure 58.

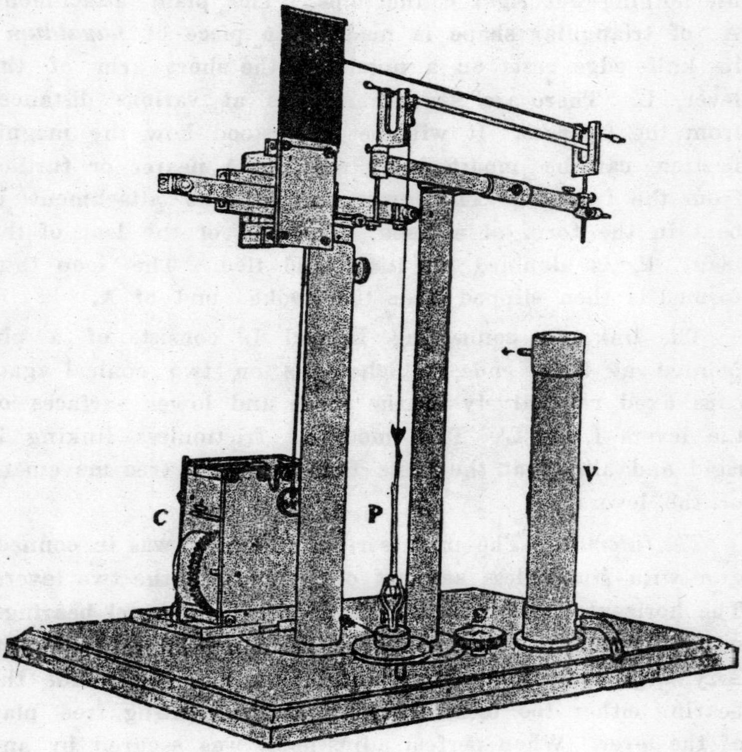

FIG. 58.—Complete apparatus. P, plant; S, micrometer screw for raising or lowering the plant; C, clockwork for periodic oscillation of plate; W, rotating wheel. V, cylindrical plant-chamber. (From a photograph.)

Connecting links.—Another puzzling difficulty lay in the fact that the magnification actually obtained was sometimes very different from the calculated value. This unreliability I was able to trace to the defects inherent in thread connections, employed at first to attach the plant to the first lever, and the first lever to the second. These flexible connections were found to undergo a variable amount of elastic yield. Hence it became necessary to use nothing but rigid connections. The plant attachment, A, of triangular shape is made of a piece of *navaldum;* its knife-edge rests on a notch at the short arm of the lever, L. There are several notches at various distances from the fulcrum. It will be understood how the magnification can be modified by moving A, nearer or further from the fulcrum. The lower end of the attachment is bent in the form of a hook. The end of the leaf of the plant P, is doubled on itself and tied. The loop thus formed is then slipped over the hooked end of A.

The link, C, connecting L and L^1 consists of a pin pointed at both ends, which rests on two conical agate cups fixed respectively to the upper and lower surfaces of the levers L and L'. This mode of frictionless linking is rigid and allows at the same time perfectly free movement of the levers.

The fulcrum.—The most serious difficulty was in connection with frictionless support of the axes of the two levers. The horizontal axis was at first supported on jewel bearings, with fine screw adjustment for securing lateral support. Any slight variation from absolute adjustment made the bearing either too loose or too tight, preventing free play of the lever. When perfect adjustment was secured by any chance, the movement of the levers became jerky after a few days. This I afterwards discovered was due to the deposit of invisible particles of dust on the bearings. These difficulties forced me to work out a very perfect and at the same time a much simpler device. The lever now carries two vertical pin-points which are supported on conical agate cups. The axis of the lever passes through the points of support. The friction of support is thus reduced to a minimum. The levers are kept in place under the constant pressure of their own weight. The excursion of the end of the recording lever, which represents magnified movement of growth, was now found to be without jerk and quite uniform.

EXPERIMENTAL ADJUSTMENTS.

The soil in a flower pot is liable to be disturbed by irrigation, and the record thus vitiated by physical disturbance. This is obviated by wrapping a piece of cloth round the root imbedded in a small quantity of soil. The lower end of the plant is held securely by a clamp. In order to subject the plant to the action of gases and vapours, or to variation of temperature it is enclosed in a glass cylinder (V) with an inlet and an outlet pipe (Fig. 58). The chamber is maintained in a humid condition by means of a sponge soaked in water. Different gases, warm or cold water vapours, may thus be introduced into the plant chamber.

Any quick growing organ of a plant will be found suitable for experiment. In order to avoid all possible disturbing action of circumnutation, it is preferable to employ either radial organs, such as flower peduncles and buds of certain flowers, or the limp leaves of various species of grasses, and the pistils of flowers. It is also advisable to select specimens in which the growth is uniform. I append a representative list of various specimens in which, under favourable conditions of season and temperature, the rates of growth may be as high as those given below :—

Peduncle of *Zephyranthes*	...	...	0·7	mm. per hour.	
Leaf of grass ...	...	...	1·10	,, ,,	,,
Pistil of *Hibiscus* flower	...	...	1·20	,, ,,	,,
Seedling of wheat ...	...	...	1·60	,, ,,	,,
Flower bud of *Crinum*	...	...	2·20	,, ,,	,,
Seedling of *Scirpus Kysoor*	...	...	3·00	,, ,,	,,

The specimen employed for experiment may be an intact plant, rooted in a flower pot. It is, however, more convenient to employ cut specimens, the exposed end being wrapped in moist cloth. The shock-effect of section passes off after several hours, and the isolated organ renews its growth in a normal manner. Among various specimens I find *S. Kysoor* to be very suitable for experiments on growth. The leaves are much stronger than those of wheat and different grasses, and can bear a considerable amount of pull without harm. Its rate of growth under favourable condition of season is considerable. Some specimens were found to have grown more than 8 cm. in the course of twenty-four hours, or more than 3 mm. per hour. This was during the rainy season in the month of August. But a month later the rate of growth fell to about 1 mm. per hour.

I shall now proceed to describe certain typical experiments which will show: (1) the extreme sensibility of the Crescograph; (2) its wide applicability in different investigations; and (3) its capability in determining with great precision the time-relations of responsive changes in the rate of growth. In describing these typical cases, I shall give detailed account of the experimental methods employed, and thus avoid repetition in accounts of subsequent experiments.

Determination of the absolute rate of growth: Experiment 51.—For the determination of the absolute rate, I shall interpret the results of a record of growth obtained with a vigorous specimen *S. Kysoor* on a stationary plate. The oscillation frequency of the plate was once in a second, and the magnification employed was ten thousand times. The magnified growth movement was so rapid that the record consists of a series of short dashes instead of dots (Fig. 59A). For securing regularity in the rate of

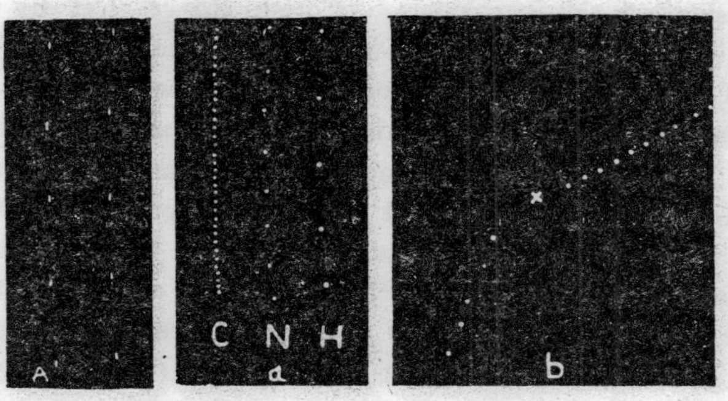

FIG. 59.—Crescographic records: (A) successive records of growth at intervals of one second (magnification 10,000 times). (*a*) Effect of temperature on a stationary plate; N, normal rate of growth; C, retarded rate under cold; H, enhanced rate under warmth: (*b*) record on moving plate, where diminished slope of curve denotes retarded rate under cold. (Magnification 2,000 times.)

growth, it is advisable that the plant should be kept in uniform darkness or in uniformly diffused light. So sensitive is the recorder that it shows a change of growth-rate due to the slight increase of illumination by the opening of an additional window. One-sided light, moreover, gives rise to disturbing phototropic curvature. With the precautions described the growth-rate in vigorous specimens is found to be very uniform.

After the completion of the first vertical series, the recording plate was moved 1 cm. to the left; the tip of the recorder was brought once more to the top by the micrometer screw, S, (Fig. 58), and the record taken once more after an interval of 15 minutes. The magnified growth for 4 seconds is 38 mm. in the first record; it is precisely the same in the record taken fifteen minutes after. The successive growth elongations at intervals of 1 second is practically the same throughout, being 9·5 mm. This uniformity in the spacings demonstrates not only the regularity of growth under constant conditions, but also the precision of the apparatus. It also shows that by keeping the external condition constant, the normal growth-rate could be maintained uniform for at least fifteen minutes. The magnified rate of growth is nearly 1 cm. per second, and since it is quite easy to measure 0·5 mm., the Crescograph enables us to magnify and record a length of 0·0005 mm., that is to say, the sixteenth part of a wave of red light. The absolute rate of growth, moreover, can be determined in a period as short as 0·05 of a second. These facts will give some idea of the great possibilities of the Crescograph for future investigations.

As the period of experiment is very greatly shortened by the method of high magnification, I shall, in the determination of the absolute rate of growth, adopt a second as the unit of time, and μ, or *micron*, as the unit of length,—the micron, being a millionth part of a metre or a thousandth part of a millimeter.

If m be the magnifying power of the compound lever and l, the average distance between successive dots in mm. at intervals of t seconds then :—

$$\text{the rate of growth} = \frac{l}{mt} \times 10^3\mu \text{ per second.}$$

In the record given $l = 9{\cdot}5$ mm.

$$m = 10{,}000.$$

$$t = 1 \text{ second.}$$

$$\text{Hence the rate of growth} = \frac{9{\cdot}5}{10{,}000} \times 10^3\mu \text{ per sec.}$$

$$= 0{\cdot}95\mu \text{ per sec.}$$

Having demonstrated the extreme sensitiveness and reliability of the apparatus, in quantitative determination, I shall next proceed to show its wide applicability for various researches relating to the influence of external agencies in modification of growth. For this two different methods are employed. In the first of these methods, the records are taken on a stationary plate : of these the record is at first taken under normal condition, the subsequent series being obtained under the given changed condition; the increase or diminution of intervals between successive dots, in the two series, at once demonstrates the stimulating or depressing nature of the changed condition.

In the second method, the record is taken on a plate moving at an uniform rate by clockwork. A curve is thus obtained, the ordinate representing growth elongation and the abscissa the time. The increment of length divided by the increment of time gives the absolute rate of growth at any part of the curve. As long as the growth is uniform, so long the slope of the curve remains constant. If a stimulating agency enhances the rate of growth, there is an immediate upward flexure in the curve; a depressing agent, on the other hand, lessens the slope of the curve.

I shall now give a few typical examples of the employment of the Crescograph for investigations on growth: the first example I shall take is the demonstration of the influence of variation of temperature.

Stationary method: Experiment 52.—The records, given in Fig. 59*a*, were taken on a stationary plate. The specimen was *S. Kysoor;* the Crescographic magnification was two thousand times, and the successive dots at intervals of 5 seconds. The middle series, N, was at the temperature of the room. The next, C, was obtained with the temperature lowered by a few degrees. Finally H was taken when the plant-chamber was warmed. It will be seen how under cooling the spaces between successive dots have become shortened, showing the diminished rate of growth. Warming, on the other hand, caused a widening of intervals between successive dots, thus demonstrating an enhancement of the rate of growth.

Calculating from the data obtained from the figure we find :—

The absolute value of the normal rate	...	0·457μ per second.
Diminished rate under cold ...	...	0·101μ ,, ,,
Enhanced rate under warmth ...	...	0 737μ ,, ,,

Moving plate method: Experiment 54.—This was carried out with a different specimen of *S. Kysoor*, the record being taken on a moving plate (Fig. 59*b*). The first part of the curve here represents the normal rate of growth. The plant was then subjected to moderate cooling, the subsequent curve with its diminished slope denotes the depression of growth. The question of influence of temperature will be treated in a subsequent Paper of the present series in much greater detail.

Precaution against physical disturbance: Experiment 54 There may be some misgiving about the employment of such high magnification: it may be thought that the accuracy of the record might be vitiated by physical disturbance, such as vibration. In physical experimentation far greater difficulties have, however, been overcome, and the problem of securing freedom from vibration is not at all formidable. The whole apparatus need only be placed on a heavy bracket screwed on the wall to ensure against mechnical disturbance. The extent to which this has been realized will be found from the inspection of the first part of the record in figure 60, taken on a moving plate. A thin dead twig was substituted for the growing plant, and the perfectly horizontal record not only demonstrated the absence of growth movement but also of all disturbance. There is an element of physical change, against which precautions have to be taken in experiments on variation of the rate of growth at different temperatures. In order to determine its character and extent, a record was taken with the dead twig, of the effect of raising the temperature of the plant-chamber through ten degrees. The record

(Fig. 60) with a magnification of two thousand times shows that there is an expansion during the rise of temperature, and that the variable period lasted for a minute, after which there was a cessation of physical movement, the record becoming once more horizontal. The obvious precautions to be taken in such a case, is to wait for several minutes for the attainment of steady temperature. The movement caused by physical change abates in a short time whereas the change of rate of growth brought about by physiological reaction is persistent.

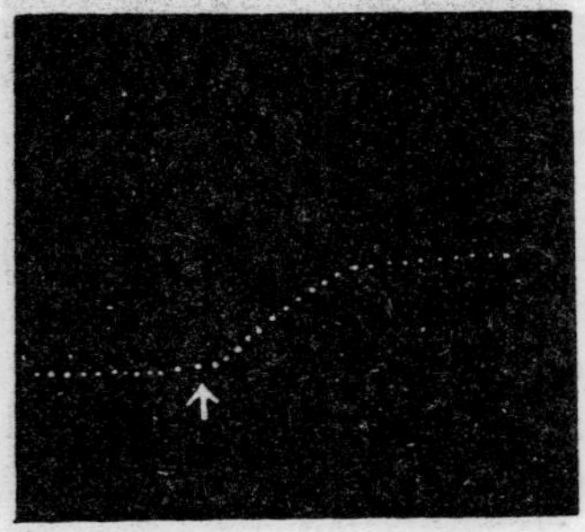

FIG. 60.—Horizontal record shows absence of growth in a dead branch; physical expansion on application of warmth at arrow followed by horizontal record on attainment of steady temperature. (Magnification 2,000 times.)

DETERMINATION OF LATENT PERIOD AND TIME-RELATIONS OF RESPONSE.

Experiment 55.—In the determination of time-relations of responsive change in growth under external stimulus, I shall take the typical case of the effect of electric shock from a secondary coil of one second's duration. Two electrodes were applied, one above and the other below the growing region of a bud of *Crinum.* The record was taken on a moving plate, the magnification employed being two thousand times, and successive dots made at intervals of two seconds. It was a matter of surprise to me to find that the growth of the plant was affected by an intensity of stimulus far below the limit of our own perception. As regards the relative sensitiveness of plant and animal, some of my experiments show that the leaf of *Mimosa pudica* in a favourable condition responds to an electric stimulus which is one-tenth the minimum intensity that causes perception in a human being. For convenience I shall designate the intensity of electric shock that is barely perceptible to us as the unit shock. When an intensity of 0·25 unit was applied to the growing organ, it responded to it by a retardation of growth. Inspection of Fig. 61 shows that there is a flexure induced in the curve in response to stimulus, the flattening of the curve denoting retardation of growth. The latent period, in this case, is 6 seconds. The normal rate was restored after 5 minutes. The intensity of shock was next raised from 0·25 unit to one unit. The second record shows that the latent period is reduced to 4 seconds, and a relatively greater retardation of growth was induced by the action of the stronger stimulus. The recovery of the normal rate was effected after the longer period of 10 minutes. I took one more record, the intensity being three units. The latent period was now reduced to 1 second, and the induced retardation was so great as to effect a temporary arrest of growth.

FIG. 61.—Time-relations of response of growing organ to electric stimulus of increasing intensities applied at the short horizontal lines. Successive dots at intervals of 2 seconds.

TABLE X.—TIME-RELATIONS OF RESPONSIVE GROWTH-VARIATION UNDER ELECTRIC SHOCK (*Crinum*).

Intensity of stimulus.	Latent period.	Normal rate.	Retarded rate.
0·25 unit.	6 seconds.	0·62 μ per sec.	0·49 μ per sec.
1 „	4 „	0·62 „	0·25 „
3 „	1 „	0·62 „	Temporary arrest of growth.

It is thus found that growth in plants is affected by an intensity of stimulus which is below human perception; that with increasing stimulus the latent period is diminished and the period of recovery increased; and that the induced retardation of growth increases continuously with the stimulus till at a critical intensity there is a temporary arrest of growth. I shall speak later of the effect induced by stimulus above this critical point.

Experiment 56.—As a further example of the capability of the Crescograph, I shall give the record of a single pulse of growth obtained with the peduncle of *Zephyranthes Sulphurea* (Fig. 62). The magnification employed was 10,000 times, the successive dots being at intervals of one second. It will be seen that the growth pulse commences with a sudden elongation, the maximum rate being 0·4 μ per sec. The pulse exhausts itself in 15 seconds, after which there is a partial recovery in the course of 13 seconds. The period of the complete pulse is 28 seconds. The resultant growth in each pulse is therefore the difference between elongation and recovery. Had a very highly magnifying arrangement not been used, the resulting rate would have appeared continuous. In other specimens, owing probably to greater frequency of pulsation and co-operation of numerous elements in growth, the rate appears to be practically uniform.

FIG. 62.—Record of a single growth-pulse of *Zephyanthes*. (Magnification 10,000 times.)

Advantages of the Crescograph.—There is no existing method which enables us to detect and measure such infinitesimal movements and their time-relations. The only attempt made in measuring minute growth has been by observing the movement of a mark on a growing plant through a microscope. The magnification available in practice is about 250 times. The observation of the movement would itself be sufficiently fatiguing. But a simultaneous estimate of the time-relations of rapidly fluctuating changes would prove so bewildering, that accurate results from this method would be altogether impossible. A $\frac{1}{12}''$ objective gives a linear enlargement of about 1,200 times. But the employment of this objective is impracticable in the measurement of growth elongation of an ordinary plant. With the Crescograph, on the other hand, we obtain a magnification which far surpasses the highest powers of a microscope, and it can be used for all plants. It does not merely detect growth but automatically records the rate of growth and its slightest fluctuation. The extreme shortness of time required for an experiment renders the study of the influence of a single factor at a time possible, the other conditions being kept constant. The Crescograph thus opens out a very extensive field of inquiry into the physiology of growth; and the discovery of several important phenomena mentioned in this Paper is to be ascribed to the extreme sensitiveness of the apparatus, and the accuracy of the method employed.

MAGNETIC AMPLIFICATION.

The magnification obtained with two levers was, as stated before, 10,000 times. It may be thought that further magnification is possible by a compound system of three levers. There is, however, a limit to the number of levers that may be employed with advantage, for the slight overweight of the last lever becomes multiplied and exerts very great tension on the plant, which interferes with the normal rate of its growth. The friction at the bearings also becomes added up by an increase in the number of levers, and this interferes with the uniformity of the movement of the last recording lever. For securing further magnification, additional material contact has, therefore, to be abandoned. I have recently been successful in devising an ideal method of magnification without contact. The movement of the lever of the Crescograph upsets a very delicately balanced magnetic system. The indicator is a reflected spot of light from a mirror carried by the deflected magnet. Taking a single lever with the lengths of two arms 125 mm. and 2·5 mm. respectively we obtain a magnification of 50 times. The magnetic system gives a further magnification of 20,000 the total magnification being thus a million times. This was verified by moving by means of a micrometer screw the short arm of the lever through 0·005 mm. The resulting deflection of the spot of light at a distance of 4 metres was found to be 5,000 mm., or a million times the movement of the short arm. It is not difficult to produce a further magnification of 50 times by attaching a second lever to the first. The total magnification would in this case be 50 million times.

A concrete idea of this will be obtained when we realise that by the Magnetic Crescograph a magnification can be obtained which is about 50,000 times greater than that produced by the highest power of a microscope. This order of magnification would lengthen a wave of sodium light to about 3,000 cm. I am not aware of any existing method by which it is possible to secure an amplification of this order of magnitude. The application of this will undoubtedly be of great help in many physical investigations, some of which I hope to complete in the near future.

Such an enormous magnification cannot be employed in ordinary investigations on growth, for the moving spot of light indicating rate of growth, passes like a flash across the screen. But it is of signal service in my investigations on growth by the Method of Balance, to be described in a future Paper. The principle of this method consists in making the spot of light, which is moving in response to growth, stationary, by subjecting the plant to a compensating movement downwards. The slightest variation caused by an external agent would make the spot of light move either to the right or to the left, according to the stimulating or depressing character of the agent. It will be understood, how extremely sensitive this method is for detection of the most minute variation in the normal rate of growth.

THE DEMONSTRATION CRESCOGRAPH.

Before proceeding with accounts of further investigations, I shall describe a form of Magnetic Crescograph with which I have been able to give before a large audience demonstration of a striking character on various phenomena of growth. The magnification obtained was so great that I had to take some trouble in reducing it. This was accomplished by the employment of a single, instead of a compound system of two levers. The reflected spot of light was thrown on a screen placed at a distance of 4 metres, and this gave a magnification of a million times; it is obvious that an increase of the distance of the screen to 8 metres would have given a magnification of 2 million times. As it was, even the lower magnification was far too great for use with quick growing plants like *Kysoor*. I, therefore, employed the slower growing flower bud of *Crinum*. It will be seen from Table X that the normal rate of growth of the lily is of the order of 0·0006 mm. per second. The normal excursion of the spot of light reflected from the Crescograph exhibiting growth was found to be 3 metres in five seconds, or 60 cm. per second. This is a million times the actual rate of growth of the *Crinum* bud. As it is easy to measure 5 mm. in the scale, it will be seen that with the Demonstration Crescograph it is possible to detect the growth of a plant for a period shorter than a hundredth part of a second.

Experiment 57.—A scale 3 metres long divided into cm. is placed against the screen. A metronome beating half seconds is started at the moment when the spot of light transits across the zero division; the number of beats is counted till the index traverses the 300 cm. At the normal temperature of the room (30 C.), the index traversed 300 cm. in five seconds. The plant chamber was next cooled to 26°C. by the blowing in of cooled water vapour; the time taken by the spot of light to traverse the scale was now 20 seconds, *i.e.*, the growth-rate was depressed to a fourth. Under continuous lowering of temperature the growth-rate became slowed down till at 21°C. there was an arrest of growth. Warm vapour was next introduced, gradually raising the temperature of the chamber to 35°C. The spot of light now rushed across the scale in a second and a half, *i.e.*, the growth was enhanced to more than three times the normal rate. The entire series of the above experiments, on the effect of temperature on growth, was thus completed in the course of 15 minutes.

SUMMARY.

A description is given of the High Magnification Crescograph, which enables an automatic record of growth magnified ten thousand times. The absolute rate of growth can be easily determined from the data given in the record.

A magnification of a million times is obtained by the employment of Magnetic amplification. An increment of growth so minute as a millionth part of a mm. or 0·00000004 inch may thus be detected. It is also possible to detect the growth of a plant for a period shorter than a hundredth part of a second.

The influence of external conditions on variation of rate of growth is obtained by two methods of record. In STATIONARY METHOD, the increase or diminution of the distance between successive dots representing magnified rate of growth, demonstrates the stimulating or depressing nature of the changed condition.

In the second, or MOVING PLATE METHOD, a curve is obtained, the ordinate representing growth elongation, and the abscissa, time. A stimulating agent causes an upward flexure of the normal curve; a depressing agent, on the other hand, lessens the slope of the curve.

The action of external stimulus induces a variation of the rate of growth, the time relations of which are found from the automatic record of the growth. The latent period is shortened with the intensity of the stimulus. A responsive variation of growth is induced by an intensity of stimulus which is below human perception.

It is often possible to obtain record of the pulsatory nature of growth-elongation. Thus with the growing peduncle of *Zephyranthes*, the growth pulse commences with a sudden elongation, the maximum rate being 0·0004 mm. per second. The pulse exhausts itself in 15 seconds, after which there is a partial recovery in course of 13 seconds, the period of complete pulse being 28 seconds. The resultant growth in each pulse is the difference between elongation and recovery.

The Magnetic Crescograph enables demonstration of principal phenomena of growth and its variation before a large audience.

XI.—EFFECT OF TEMPERATURE ON GROWTH

By

SIR J. C. BOSE,

Assisted by

SURENDRA CHUNDER DASS, M.A.

Accurate determination of the effect of temperature on growth presents many serious difficulties on account of numerous complicating factors. In nature, the upper part of the plant is exposed to the temperature of the air, while the root underground is at a very different temperature. Growth, we shall find, is modified to a certain extent by the ascent of sap. (See p. 189, *Expt. 69.*) The activity of this latter process is determined by the temperature to which the roots are subjected. The difficulty may be removed to a certain extent by placing the plant in a thermal chamber, with arrangement for regulating the temperature of the air. The air is a bad conductor of heat, and there is some uncertainty of the interior of the plant attaining the temperature of the surrounding air, unless the plant is long exposed to the definite and constant temperature of the plant chamber. Observation of the effects of different temperatures then becomes a prolonged process, with the possibility of vitiation of results by autonomous variation of growth. Reduction of the period of experiment by rapidly raising the temperature of the chamber introduces fresh difficulties; for a sudden variation of temperature often acts like an excitatory shock. This drawback may to some extent be obviated by ensuring a gradual change of temperature. This is by no means an easy process, for even with care the rise of temperature of the air cannot be made perfectly uniform, and any slight irregularity gives rise to sudden fluctuations in the magnified record of growth. Another difficulty arises from the radiation of heat-rays from the sides of the thermal chamber. These rays, I shall in a different Paper show, induce a retardation of growth. The effect of rise of temperature in acceleration of growth is thus antagonised by the action of thermal radiation. This trouble may be minimised by having the inner surface of the thermal chamber of bright polished metal, since the radiating power of a polished surface is relatively feeble.

The contrivance which I employ for ensuring a gradual rise of temperature, consists of a double-walled cylindrical metallic vessel; the plant is placed in the inner chamber, the walls of which are coated with electrically deposited silver and polished afterwards, and at the bottom of which there is a little water. The space between the inner and outer cylinder is filled with water, in which is immersed a coiled copper pipe. Hot water from a small boiler enters the inlet of the coiled pipe and passes through the outlet at the lower end. The water in the outer cylinder is thus gradually raised by flow of hot water in the coiled pipe. The rate of flow of hot water, on which the rate of rise of temperature depends, is regulated by a stop-cock. The air of the inner chamber in which the plant is placed, may thus be adjusted for a definite temperature. The small quantity of water in the inner chamber keeps its air in a humid condition, since dry hot air by causing dessication interferes with normal growth.

METHOD OF DISCONTINUOUS OBSERVATION.

Experiment 58.—High magnification records are taken for successive periods of ten seconds, for selected temperatures, maintained constant during the particular observation. In figure 63 is given records of rate of growth obtained with a specimen of *Kysoor* at certain selected temperatures. It will be seen that the rate of growth increases with the rise of temperature to an optimum, beyond which the growth-rate undergoes a depression. In the present case the optimum temperature is in the neighbourhood of 35°C.

FIG. 63.—Effect of temperature on growth, and determination of optimum temperature.

METHOD OF CONTINUOUS OBSERVATION.

The method of observation that I have described above is not ideally perfect, but the best that could be devised under the circumstances. A very troublesome complication of pulsations in growth, arises at high temperatures, which render further record extremely difficult. Growth is undoubtedly a pulsatory phenomenon; but under favourable circumstances these merge practically into a continuous average rate of elongation. At a high temperature the effect of certain disturbing factors comes into prominence. This may be due to some slight fluctuation in the temperature of the chamber, or to the effect of thermal radiation from the side of the chamber. This disturbing influence is most noticed at about 45°C, rendering the record of growth above this point a matter of great uncertainty. It will presently be shown that in plants immersed in water-bath growth is often found to persist even up to 57°C.

The only way of removing the complication arising from thermal radiation lies in varying the temperature condition of the plant, by direct contact with water at different temperatures. This procedure will also remove uncertainty regarding the body of the plant assuming the temperature of surrounding non-conducting air. The disturbing effect of sudden variation of temperature is also obviated by a more uniform regulation of rise of temperature. The inner cylinder containing the plant is filled with water; heat from gradually warmed water in the outer cylinder is conducted across the inner cylinder made of thin copper and raises the temperature of the water contained in the inner cylinder with great uniformity. A clock-hand goes round once in a minute; the experimenter, keeping his hand on the stop-cock, adjusts the rate of rise of water in the inner cylinder, so that there is a rise, say, of one-tenth of a degree every 6 seconds or of one degree every minute. The mass of water acts as a governor, and prevents any sudden fluctuations of temperature. The adoption of this particular device eliminated the erratic changes in the rate of growth that had hitherto proved so baffling.

The elongation recorded by the Crescograph will now be made up of (1) physical expansion, (2) expansion brought about by absorption of water, and (3) the pure acceleration of growth. The disentanglement of these different elements presented many difficulties. I was, however, able to find out the relative values of the first two factors in reference to the elongation of growth. This was done by carrying out a preliminary experiment with a specimen of plant in which growth had been completed. It was raised through 20°C in temperature, records being taken both at the beginning and at the end. This was for obtaining a measure of the physical change due to temperature, and also of the change brought about by absorption of water. I should state here that for the method of continuous record of growth which I contemplated, the record had to be taken for about 18 minutes. The magnification had to be lowered to 250 times to keep the record within the plate. With this magnification, the fully grown specimen did not show in the record a change even of 1 mm. in length in 18 minutes, while the growing plant under similar circumstances exhibited an elongation of 100 mm., or more. In records taken with low magnification, the effect of physical change is quite negligible.

DETERMINATION OF THE CARDINAL POINTS OF GROWTH.

The cardinal points of growth are not the same in different plants; they are modified in the same species by the climate to which the plants are habituated; the results obtained in the tropics may thus be different from those obtained in colder climates. At the time of the experiment, the prevailing temperature at Calcutta in day time was about 30°C.

Temperature minimum: Experiment 59.—For the determination of the minimum, I took a specimen of *S. Kysoor*, and subjected it to a continuous lowering of temperature, by regular flow of ice-cold water in the outer vessel of the plant-chamber. Record was taken on a moving plate for every degree fall of temperature; growth was found to be continuously depressed, till an arrest of growth took place at 22°C (Fig. 64).

The arrested growth was feebly revived at 23 C, after which with further rise of temperature there was increased acceleration. The optimum point was reached at about 34°C.

In some plants the optimum is reached at about 28 C, and the rate remains constant for the next 10 degrees or more.

FIG. 64.

FIG. 65.

FIG. 64.—Record of effect of fall of temperature from 30°C to arrest of growth at 22°C.

FIG. 65.—Effect of rise of temperature from 53°C to 60°C. A sudden contraction, indicative of death-spasm, takes place at 60°C.

Temperature maximum: Experiment 60.—For the determination of the maximum, the temperature was raised much higher. At 55 C. growth was found to be greatly retarded with practical arrest at 58 C. At 60 C there occurred a sudden spasmodic contraction (Fig. 65), which I have own elsewhere to be the spasm of death. This mechanical asm at 60 C is also strikingly shown by various pulvinate organs. An electric spasm of galvanometric negativity, and a sudden diminution of electrical resistance also take lace at the critical temperature of 60°C.*

I have described the immediate effect at the critical point. Long maintenance at a temperature few degrees

* Bose—"Plant Response," p. 168; "Comparative Electro-Physiology," p. 202, p. 546.

below 60°C, will no doubt be attended with the death of the organ. Fatigue is also found to lower the death-point.

THE THERMO-CRESCENT CURVE.

Experiment 66.—I was next desirous of devising a method by which an automatic and continuous record of the plant should enable us to obtain a curve, which would give the rate of growth at any temperature, from the arrested growth at the minimum to a temperature as high as 40°C. In order to eliminate the elements of spontaneous variation, the entire record had to be completed within a reasonable length of time, say about 18 minutes for a rise of as many degrees in temperature. This gives a rate of rise of 1°C. for one minute. Separate experiments showed that at this rate of *continuous* rise of temperature there is practically no lag in the temperature assumed by thin specimens of plants. For observation during a limited range I use the slower rate of rise at 1°C per two minutes. But the result obtained by slower rise was found not to differ from that obtained with one degree rise per minute. The curve of growth is taken on a moving plate, which travels 5mm. per minute. Successive dots are made by the recording lever at intervals of a minute during which the rise of temperature is 1°C. A *Thermo-crescent Curve* is thus obtained, the ordinate of which represents increment of growth, and the abscissa, the time. As the temperature is made to rise one degree per minute, the abscissa also represents rise of temperature (Fig. 66). The vertical distance between two successive dots thus gives increment of growth in one minute for 1 degree rise of temperature from T to T'. If l represents this length, t the interval of time (here 60 sec.), and m the magnifying power of the recorder, then the rate of growth for the mean temperature $\frac{T+T'}{2}$ is found from the formula : rate of growth at $\frac{T+T'}{2} = \frac{l}{m \times t \times 60} 10^3\mu$ per sec.

22°C 30°C 40°C

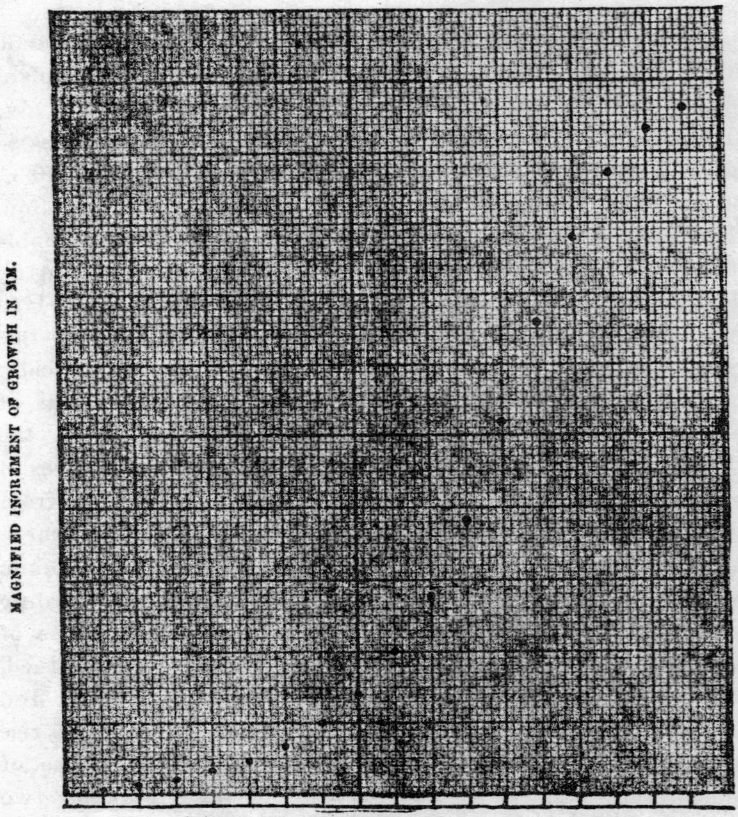

FIG. 66.—The Thermo-crescent Curve.

TABLE XI.—RATE OF GROWTH FOR DIFFERENT TEMPERATURES.

Temperature.	Growth.	Temperature.	Growth.
22°C ...	0·00 μ per sec.	31°C ...	0·45 μ per sec.
23°C ...	0·02 μ " "	32°C ...	0·60 μ " "
24°C ...	0·04 μ " "	33°C ...	0·80 μ " "
25°C ...	0·06 μ " "	34°C ...	0·92 μ " "
26°C ...	0·08 μ " "	35°C ...	0·84 μ " "
27°C ...	0·12 μ " "	36°C ...	0·64 μ " "
28°C ...	0·16 μ " "	37°C ...	0·48 μ " "
29°C ...	0·22 μ " "	38°C ...	0·30 μ " "
30°C ...	0·32 μ " "	39°C ...	0·16 μ " "

I give in figure 67 a curve showing the relation between temperature and growth.

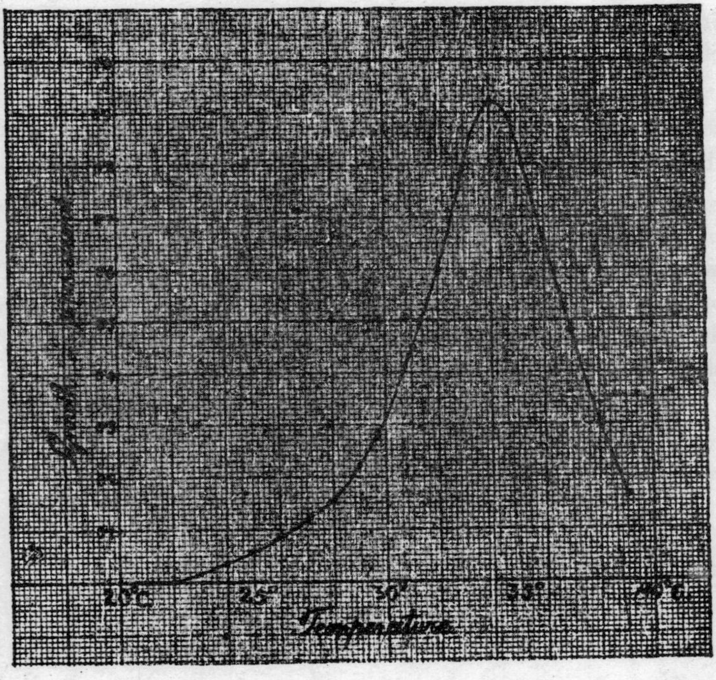

FIG. 67.—Curve showing relation between temperature and rate of growth.

It will thus be seen that, in the course of an experiment lasting about twenty minutes, data have been obtained which enable us to determine the rates of growth through a wide range of temperature. We have likewise been able by the first method to make very accurate determinations of the temperature maximum and minimum. In short, by adopting the methods described, the cardinal points of growth and the rate of growth at any temperature, may be determined with a precision unattainable by the older methods, of averages or of prolonged observation.

SUMMARY.

Temperature induces variation in the rate of growth. In accurate determination of the growth, the disturbing effect of radiation of heat has not be eliminated.

A continuous record of growth under uniform rise of temperature gives the Thermo-crescent curve, from which the rate of growth at any temperature may be deduced.

Different plant-tissues exhibit characteristic differences in their cardinal points of growth. In *Kysoor*, growth is arrested at the temperature minimum of 22°C. The optimum temperature is at 34°C., after which growth-rate declines and becomes completely arrested at 58°C. At' 60°C. there is a sudden spasmodic contraction of death.

In other plants the cardinal points are different. In some plants the optimum growth is attained at 28°C. and remains constant up to 38°C.

XII.—THE EFFECT OF CHEMICAL AGENTS ON GROWTH

By

SIR J. C. BOSE,

Assisted by

GURUPRASANNA DAS.

Chemical agents are found to exert characteristic actions on growth. The method of investigation sketched here opens out an extended field of investigation. The effect of a chemical substance, I find, to be modified by (1) the strength of the solution, (2) the duration of application, and (3) the condition of the tissue. A poisonous substance in minute doses is often found to exert a stimulating action. Too long continued action of a stimulant, on the other hand, exerts a depressing effect. The influence of the tonic condition is shown by the fact that while a given dilution of a poisonous substance kills a weak specimen, the same poisonous solution, applied to a vigorous specimen, actually stimulates and enhances the rate of the growth. I give below descriptions of a few typical reactions.

The reagent, when in a liquid form, is locally applied on the growing organ. The records, taken before and after the application, exhibit the stimulatory or depressing character of the reagent. A different method of application of the reagent is employed for plants with extended region of growth. The specimen is then enclosed in a glass cylinder, with inlet and outlet pipes. The cylinder is first filled with water, and the normal rate of growth recorded. This rate remains constant for several hours; but prevention of access of air for too long a time affects the normal growth. After obtaining normal record, water charged with the giving chemical agent is passed into the cylinder; and the subsequent record shows the characteristic effect of the reagent. The introduction of a gas into the chamber offers no difficulty.

EFFECT OF STIMULANTS.

Hydrogen Peroxide: Experiment 62.—This reagent, as supplied by Messrs. Parke Davis & Co., was diluted to 1 per cent. and applied to the growing plant. Its stimulating action on growth is demonstrated in the right hand record of Fig. 68*a*, where the rate of growth is seen enhanced two and a half times the normal rate.

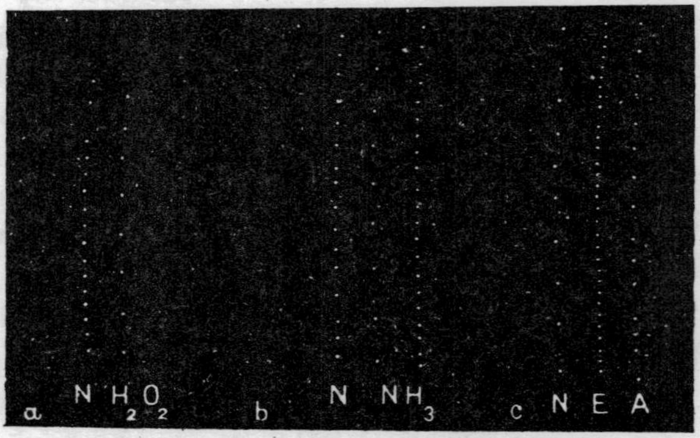

FIG. 68.—Effect of chemical agents: (*a*) Acceleration of growth under H_2O_2, (*b*) Effect of NH_3, preliminary acceleration followed by retardation. (*c*) Effect of ether (E) and recovery (A).

Ammonia: Experiment 63.—The immediate effect of dilute vapour of this reagent is an enhancement of growth, seen in the middle record of Fig. 68*b*, where the rate is seen to be double the normal. Continued action, however, caused a depression; the third record of this series shows this, where the reduction is three-fourths of the normal rate.

EFFECT OF ANÆSTHETICS.

Ether : Experiment 64.—In Fig. 68*c*, the records exhibit the effect of introduction of ether vapour into the plant chamber, and its recovery after the removal of the vapour. Ether is seen to depress the rate of growth to a little more than a third of the normal rate. The recovery is seen to be nearly complete half an hour after the removal of the vapour.

Carbonic Acid: Experiment 65.—The action of this gas is very remarkable. The plant was immersed in water and normal record taken; the plant chamber was now filled with water, charged with carbonic acid gas. This induced a very marked acceleration of growth (Fig. 69). In a seedling of Onion, the increase was found to be two and a half times. In the flower bud of *Crinum*, the rate was found enhanced threefold from the normal 0·25 μ to 0·75 μ per second. After this preliminary enhancement, there was a depression of growth within 15 minutes of the application, the

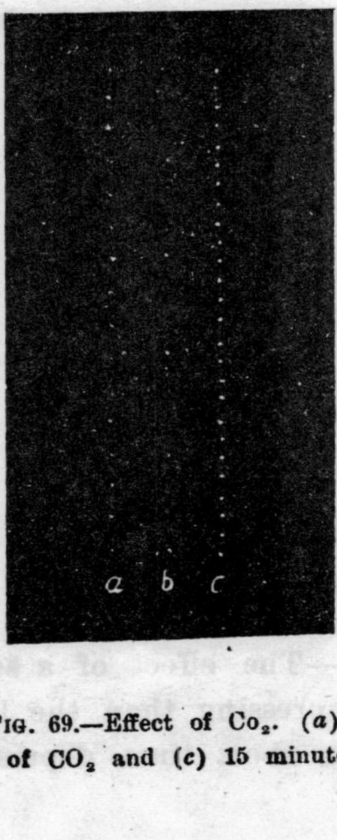

FIG. 69.—Effect of CO_2. (*a*) Normal record; (*b*) immediately after application of CO_2 and (*c*) 15 minutes after.

rate being now reduced to 0·15 μ per second. These effects were found to take place equally in light or in darkness.

ACTION OF DIFFERENT GASES.

Coal Gas: Experiment 66.—Coal gas induces a depression. It is curious that subjection to the action of this gas does not produce so evil an effect as one would expect. The introduction of the gas had reduced the growth-rate to more than half; but there was a recovery half an hour after the introduction of fresh air.

Sulphuretted Hydrogen: Experiment 67.—This gas not only exerts a depressing effect, but its after-effect is also very persistent. The plant experimented on was very vigorous and its rate of growth was depressed to half by subjection to the action of the gas for a short time. The record taken half an hour after the introduction of fresh air did not exhibit any recovery.

ACTION OF POISONS.

Ammonium Sulphide: Experiment 68.—This reagent in dilute solution retards growth, and in stronger solution acts as a poison. The following results were obtained with a wheat seedling under different strengths of solution:—

Normal rate	...	0·30 μ per sec.
0·5 per cent. solution	...	0·15 μ " "
2·0 " " "	...	0·08 μ " "

Copper Sulphate: Experiment 69.—The effect of a solution of this reagent is far more depressing than the last. One per cent. solution acting for a short time depressed the rate from 0·45 μ to 0·13 μ per sec. Long continued action of the poisonous solution kills the plant.

SUMMARY.

The effect of a chemical agent is modified by the strength of the solution, the duration of application and the tonic condition of the tissue.

Dilute solution of hydrogen peroxide induces an acceleration of growth.

The action of dilute vapour of ammonia is a preliminary enhancement followed by depression of growth.

Ether vapour depresses the rate of growth. On the removal of the vapour there is a recovery of the normal rate.

The effect of carbonic acid is a great enhancement of the rate of growth; after this preliminary action, growth undergoes a decline. The effect described takes place equally in light or in darkness.

Coal gas induces a depression of the rate of growth from which there is a recovery after the removal of the gas. The action of sulphuretted hydrogen is far more toxic, the after-effect being very persistent.

Solution of ammonium sulphide induces increasing retardation of growth, with the strength of the solution. Copper sulphate solution acts as a toxic agent, retarding the rate of growth and ultimately killing the plant.

XIII.—EFFECT OF VARIATION OF TURGOR AND OF TENSION ON GROWTH

By

SIR J. C. BOSE.

The movements of leaves of sensitive plants are caused by variation of turgor in the pulvinus induced by stimulus. The down movement or *negative* response of *Mimosa* is caused by a diminution or negative variation of turgor, while the erection or *positive* response is brought about by an increase, or positive variation of turgor.

We shall now investigate the change induced in a growing organ in the rate of growth by variation of turgor. Turgor may be increased by enhancing the rate of ascent of sap or by an artificial increase of internal hydrostatic pressure. A diminution of turgor may, on the other hand, be produced by withdrawal of water through plasmolysis. In order to maintain a constant terminology I shall designate an increase, as the positive, and a diminution, the negative variation of turgor.

RESPONSE TO POSITIVE VARIATION OF TURGOR.

In experimenting with *Mimosa* the plant was subjected to the condition of drought, water being withheld for a day. On supplying water, the leaf, after a short period, exhibited a positive or erectile movement (*Expt. 12*). The delay was evidently due to the time taken by the water absorbed by root to reach the responding organ.

Method of Irrigation: Experiment 70.—In order to investigate the effect of enhanced turgor on growth, I took a specimen of

Kysoor which had been dug up with an attached quantity of soil; this latter was enclosed in a small bag. The plant was then securely clamped and fixed on a stand. This precaution was taken to prevent upward displacement by the swelling of the soil in flower pot of the plant under irrigation. The specimen was then subjected to a condition of drought, water being withheld for a day. The depressed rate of growth is seen in record (Fig. 70). Ordinary cold water was now applied at the root, the effect of which is seen in record C. Finally the record (H) was obtained after irrigation with tepid water. It will be seen that the spaces between successive dots, representing magnified growth at intervals of ten seconds, are very different. While a given elongation took place under drought in 19×10 seconds, a similar lengthening took place, after irrigation with cold water, in 13×10 seconds, and after irrigation with warm water in 3×10 seconds. Irrigation with warm water is thus seen to increase the rate of growth more than six times.

FIG. 70.—Effect of irrigation: D, record of growth under drought; C. acceleration after irrigation with cold water; H, enhanced acceleration on irrigation with warm water. (*S. Kysoor.*)

The enhancement of the rate of growth on irrigation with cold water took place after seventy seconds. The interval will obviously depend on the distance between the root by which the water is absorbed and the region of growth. It will further depend on the activity of the process of the ascent of sap. The time interval is greatly reduced when this activity is in any way increased. Thus the responsive growth elongation after application of warm water was very much quicker; in the case described it was less than 20 seconds. With regard to application of warm water, the variation of temperature should not be too sudden ; it should commence with tepid, and end with warm water. *Sudden* application of hot water brings about certain complications due to excitatory effect. As regards the persistence of after-effect of a single application of warm water, it should be remembered that the absorbed water gradually cools down. In an experiment with a peduncle of *Zephyranthes* the growth under partial drought was found to be 0·04 μ per second ; application of warm water increased the growth rate to 0·20 μ per second. After 15 minutes the growth rate fell to 0·13 μ per second ; and after an hour to 0·08 μ per second. It will be noted that even then the rate was twice the initial rate before irrigation.

TABLE XII.—EFFECT OF IRRIGATION.

Specimen.	Condition of Experiment.	Rate of growth.
Kysoor ...	Dry soil ...	0·21 μ per second.
	Irrigation with cold water	0·30 μ „ „
	Irrigation with warm water	1·33 μ „ „
Peduncle of *Zephyranthes.*	Dry soil	0·04 μ „ „
	Irrigation with warm water	0·20 μ „ „

EFFECT OF ARTIFICIAL INCREASE OF INTERNAL HYDROSTATIC PRESSURE.

Increased turgor was, next, artificially induced by increase of internal hydrostatic pressure.

Experiment 71.—The plant was mounted water-tight in the short limb of an U-tube, and subjected to increased hydrostatic pressure by increasing the height of the water in the longer limb. Table XIII shows how increasing pressure enhances the rate of growth till a critical point is reached, beyond which there is a depression. This critical point varies in different plants.

TABLE XIII.—EFFECT OF INCREASED INTERNAL HYDROSTATIC PRESSURE (*Kysoor*).

Specimen.	Hydrostatic pressure.	Rate of growth.
	Normal	0·18 μ per second.
No. I ...	2 cm. pressure ...	0·20 μ " "
	4 cm. " ...	0·11 μ " "
	Normal	0·13 μ " "
No. II ...	1 cm. pressure ...	0·20 μ " "
	3 cm. " ...	0·18 μ " "
	4 cm. " ...	0·15 μ " "

RESPONSE TO NEGATIVE VARIATION OF TURGOR.

I shall now describe the influence of induced diminution of turgor on the rate of growth.

Method of plasmolysis: Experiment 72.—Being desirous of demonstrating the responsive growth variations of opposite signs in an identical specimen under alternate increase and diminution of turgor, I continued the experiment with the same peduncle of *Zephyranthes* in which the growth acceleration was induced by irrigation with warm water. In that experiment the growth rate of 0·04 μ per second was enhanced to 0·20 μ per second after irrigation. A strong solution of KNO_3 was now applied at the root;

FIG. 71.—Effect of alternate increase and diminution of turgor on the same specimen : N, normal rate under drought ; H, enhanced rate under irrigation with warm water ; N′, normal permanent rate after irrigation ; P, diminished rate after plasmolysis (*Zephyranthes*).

and the growth-rate fell almost immediately to 0·03 μ per second, or nearly to one-third the previous rate, the depression induced being thus greater than under condition of drought (Fig. 71).

TABLE XIV.—EFFECT OF ALTERNATE VARIATION OF TURGOR ON GROWTH (*Zephyranthes*).

Condition of Experiment.	Rate of growth.
Dry soil	0·04 μ per second.
Application of warm water ...	0·20 μ ,, ,,
Steady growth after 1 hour ...	0·08 μ ,, ,,
Application of KNO_3 solution ...	0·03 μ ,, ,,

From the series of results that have been given above, it will be seen that employing very different methods of turgor variation, the rate of growth, within limits, is enhanced by an increase of turgor. A diminution or negative variation of turgor, on the other hand, brings about a retardation or negative variation in the rate of growth. We should, in this connection, bear in mind the fact that, growth is dependent on protoplasmic activity, and the variation of turgor itself is also determined by that activity.

RESPONSE OF MOTILE AND GROWING ORGANS TO VARIATION OF TURGOR.

I have already described (p. 40) the effects of variation of turgor on the motile pulvinus of *Mimosa*. There is a strict correspondence between the responsive movement of the leaf of *Mimosa* and the movement due to growth, which is summarized as follows:—

(1) *An increase or positive variation of turgor induces an erection or positive response of the leaf of* Mimosa, *and a positive variation or enhancement of the rate of growth.*

(2) *A diminution or negative variation of turgor induces a fall or negative response of the leaf of* Mimosa, *and a negative variation or retardation of the rate of growth.*

EFFECT OF EXTERNAL TENSION.

Experiment 73.—The recording levers are at first so balanced that very little tension is exerted on the plant. Record of normal growth is taken of a specimen of *Crinum*. The tension is gradually increased from one gram to ten grams. The table given above shows how growth-rate increases with the tension till a limit is reached, after which there is a retardation.

TABLE XV.—EFFECT OF TENSION ON GROWTH.

Tension.	Rate of growth.
0 (Normal)	0·41 μ per second.
4 grams	0·44 μ ,, ,,
6 ,, ...	0·48 μ ,, ,,
8 ,, ...	0·52 μ ,, ,,
10 ,, ...	0·40 μ ,, ,,

SUMMARY.

Increase of turgor induced by irrigation enhances the rate of growth. Irrigation with warm water induces a further augmentation of the rate of growth.

The latent period for enhancement of growth depends on the distance of growing region from the root. The latent period is reduced when the plant is irrigated with warm water.

Artificial increase of internal hydrostatic pressure, up to a critical degree, enhances the rate of growth.

A diminution or negative variation of turgor depresses the rate of growth.

There is a strict correspondence between the responsive movement of the leaf of *Mimosa*, and the movement due to growth. An increase or positive variation of turgor induces an erection of positive response of the leaf of *Mimosa*, and a positive variation or enhancement of the rate of growth. A diminution or negative variation of turgor induces a fall or negative response of the leaf of *Mimosa*, and a negative variation or retardation of the rate of growth.

External tension within limits, enhances the rate of growth.

XIV.—EFFECT OF ELECTRIC STIMULUS ON GROWTH

By

SIR J. C. BOSE,

Assisted by

GURUPRASANNA DAS.

In plant physiology, the word 'stimulus' is often used in a very indefinite manner. This is probably due to the different meanings which have been attached to the word. An agent is said to *stimulate* growth, when it induces an acceleration. But the normal effect of stimulus is to cause a retardation of growth. It is probably on account of lack of precision in the use of the term that we often find it stated, that a stimulus sometimes accelerates, and at other times, retards growth. In order to avoid any ambiguity, it is very desirable that the term stimulus should always be used in the sense as definite as in animal physiology. An induction shock, a condenser discharge, the make or break of a constant current, a sudden variation of temperature, and a mechanical shock bring about an excitatory contraction in a muscle. These various forms of stimuli cause, as we have seen, a similar excitatory contraction of the motile pulvinus of *Mimosa pudica*. We shall enquire whether the diverse forms of stimuli enumerated above, exert similar or different reactions on the growing organ.

EFFECT OF ELECTRIC STIMULUS OF VARYING INTENSITY AND DURATION.

The form of stimulus which is extensively used in physiological investigations, is the electric stimulus of induction shock which is easily graduated by the use of the well known sliding induction coil, in which the approach of the secondary to the primary coil, indicated by the higher reading of the scale, gives rise to increasing intensity of stimulus. The retarding effect of electrical stimulus on growth has already been demonstrated in record taken on a moving plate (Fig. 61).

I shall adopt for unit stimulus, that intensity of electric shock which induces a barely perceptible sensation in a human being. It is very interesting to find, as stated before, that growth is often affected by an electric stimulus, which is below the range of human perception.

Effect of Intensity: Experiment 74.—I shall now describe a typical experiment on the effect of intensity of

FIG. 72.—Effects of electrical stimulus of increasing intensities: of 0·25 unit, 1 unit, and 3 units. Short dashes represent the moments of application of stimulus.

stimulus in retarding the rate of growth. The normal rate of growth of the bud of *Crinum* was 0·35 μ per second. On the application of electric shock of unit intensity for 5 seconds, the rate became reduced to 0·22 μ per second. When the stimulus was increased to 2 units, the retarded rate of growth was 0·07 μ per second. When the intensity was raised to 4 units, there was a complete arrest of growth. In figure 72 is given records of a different experiment which show the effects of increasing intensity of stimulus in retardation of growth.

Effect of continuous stimulation: Experiment 75.—The effect of continuous stimulation of increasing intensity will be seen in the record (Fig. 73), taken on a

FIG. 73.—Effect of continuous electric stimulation of increasing intensity. The last record exhibits the actual shortening of the growing organ under strong stimulus.

moving plate. On application of continuous stimulus of increasing intensity an increased flexture was produced in the curve, which denoted greater retardation in the rate of growth. When the intensity of stimulus was raised to 3 units, there was induced an actual contraction.

CONTINUITY BETWEEN INCIPIENT AND ACTUAL CONTRACTION.

It will thus be seen, that external stimulus of electric shock induces a reaction which is of opposite sign to the normal growth elongation or expansion. We may conveniently describe this effect as 'incipient' contraction; for under increasing intensity of stimulus, the contractile reaction, opposing growth elongation, becomes more and more pronounced; at an intermediate stage this results in an arrest of growth; at the further stage, it culminates in an actual shortening of the organ. There is no break of continuity in all these stages. I shall, therefore, use the term 'contraction' in a wider sense, including the 'incipient' which finds expression in a retardation of growth.

In Table XVI is given the results of certain typical experiments on the effect of stimulus of increasing intensity and duration.

TABLE XVI.—EFFECT OF INTENSITY AND DURATION OF ELECTRIC STIMULUS ON GROWTH.

Duration of Application.	Intensity.	Rate of growth.
	Normal ...	0·35 μ per second.
5 seconds ...	1 unit ...	0·22 μ „ „
„ ...	2 units ...	0·07 μ „ „
„ ...	4 „ ...	Arrest of growth.
	Normal ...	0·30 μ per second.
Continuous stimulation	0·5 unit ...	0·20 μ „ „
„	1 „ ...	0·09 μ „ „
„	3 „ ...	Contraction.

With regard to the question of immediate and after-effect of stimulus, I find great difficulty in drawing a line of demarcation. Owing to physiological inertia there is a delay between the application of stimulus and the initiation of responsive reaction (latent period); owing to the same inertia, the physiological reaction is continued even on the cessation of stimulus. All responsive reactions are thus after-effects in reality. The latent period is shortened under strong stimulus, but the contractile reaction becomes more persistent. When the stimulus is moderate or feeble, the recovery from incipient contraction takes place within a short time. Stimulus, under certain circumstances, is found to improve the 'tone' of the tissue, and as we shall presently see bring about, as the after-effect, an enhancement of the rate of growth.

The effect of electric stimulus is thus an incipient or actual contraction.

SUMMARY.

In normal conditions electric stimulus induces an incipient contraction exhibited by the retardation of the rate of growth. Growth is often affected by an electric stimulus which is below human perception.

Under increasing intensity of stimulus, the contractile reaction opposing growth elongation becomes more and more pronounced. At a critical intensity of stimulus growth becomes arrested. Under stronger intensity of stimulus growing organ undergoes an actual shortening in length.

There is continuity between the incipient contraction seen in retardation, arrest of growth, and contraction of the organ under stronger stimulus.

The latent period of responsive variation of growth is shortened under stronger stimulus, but the period of recovery becomes protracted.

XV.—EFFECT OF MECHANICAL STIMULUS ON GROWTH

By

SIR J. C. BOSE.

Amongst the various stimuli which induce excitation in *Mimosa* may be mentioned the irritation caused by rough contact, by prick, or wound. Friction causes moderate stimulation, from which the excitated pulvinus recovers within a short time. But a prick or a cut induces a far more intense and persistent excitation; the recovery becomes protracted, and the wounded pulvinus remains contracted for a long period.

I shall now describe the effect of mechanical irritation on growth. For moderate stimulus, I employ rough contact or friction; more intense stimulation is caused by a prick or a cut.

EFFECT OF MECHANICAL IRRITATION.

Experiment 76.—In this experiment, I took a peduncle of *Zephyranthes*, which had a normal rate of growth of 0·18 μ per second. I then caused mechanical irritation by rubbing the surface with a piece of card-board. The mechanical stimulation was found to have caused a retardation of growth, the depressed rate being 0·11 μ per second, or three-fifths the normal rate. As this particular mode of stimulation was very moderate, the normal of rate growth was found to be restored after a short period of rest. After 15 minutes, the rate became 0·14 μ per second; after an hour the recovery was complete, the rate being now 0·18 μ per second, the normal rate before stimulation (Fig. 74*a*). We shall presently see that not only is the growth rate greatly depressed under intense stimulation, but the period of recovery also becomes very much protracted.

I have often been puzzled by the fact, that specimens apparently vigorous exhibited little or no growth, after attachment to the recorder. After waiting in vain for an hour, I had to discard them for others with equally unsatisfactory results. One of these specimens happened to be left attached to the recorder overnight, and I was surprised to find that the specimen, which had shown no growth the

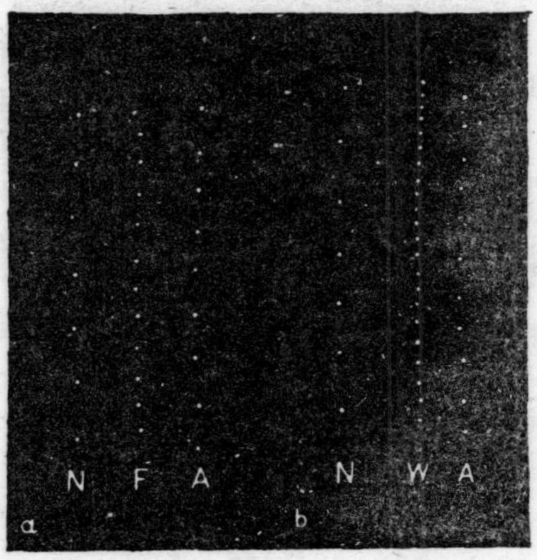

FIG. 74.—(*a*) N, normal rate of growth; F, retarded rate immediately after friction; A, partial recovery after 15 minutes.
(*b*) N, normal; W, immediately after wound; C, an hour after. (Successive dots at intervals of 5″.)

previous evening, was now exhibiting vigorous growth after being left to itself for 12 hours. I then realised that the temporary abolition of growth must have been due to the irritation of somewhat rough handling during the process of mounting and attachment of the specimen to the recorder.

In the matter of mechanical stimulation, some specimens are more irritable than others. The persistence of after-effect of irritation in retardation of growth will be demonstrated in the following experiments, where the stimulus employed was more intense.

EFFECT OF WOUND.

A prick causes an intense excitation in *Mimosa*. I tried the effect of this form of stimulation on responsive variation in growth.

Experiment 77.—The specimen was the same as had been employed in the last experiment. After moderate stimulation due to friction it had, in the course of an hour, completely recovered its normal rate of growth of 0·18 μ per second. I now applied the stimulus of pin prick ; the actual injury to the tissue due to this was relatively slight ; but the retardation of growth induced by this more intense mode of stimulation was very great. With moderate mechanical friction the rate had fallen from 0·18 μ to 0·11 μ per second, *i.e.*, to three-fifths the normal rate: in consequence of prick the depression was from 0·18 μ to 0·05 μ per second, *i.e.*, to less than a third of the normal rate. After 15 minutes the rate recovered from 0·05 μ to 0·07 μ per second. After moderate friction the recovery was complete after an hour ; but in this case the recovery after an equal interval was only three-fourths of the original, the rate being now 0·12 μ per second (Fig. 33b). I next applied the more intense stimulus caused by a longitudinal cut,

This caused a depression of growth rate to 0·04 μ per second. A transverse cut, I find, gives rise to a more intense stimulation, than a longitudinal slit.

TABLE XVII.—EFFECT OF MECHANICAL IRRIGATION AND OF WOUND ON GROWTH.

(*Zephyranthes.*)

Nature of stimulus.	Condition.	Rate of growth.
Mechanical friction.	Normal rate ...	0·18 μ per sec.
	Immediately after stimulation ...	0·11 μ ,,
	15 minutes after stimulation ...	0·14 μ ,,
	60 minutes after stimulation ...	0·18 μ ,,
Prick with needle	Normal rate ...	0·18 μ per sec.
	Immediately after stimulation ...	0·05 μ ,,
	15 minutes after stimulation ...	0·07 μ ,,
	60 minutes after stimulation ...	0·12 μ ,,

The effect of mechanical stimulus on growth is thus similar to that induced by electrical stimulus. Moderate stimulus of rough contact induces an incipient contraction, seen in retardation of growth, the recovery being complete in the course of an hour ; but intense stimulation, induced by wound, gives rise to greater and more persistent retardation of growth.

SUMMARY.

Mechanical stimulus induces incipient contraction or retardation of rate of growth, the effect being similar to that induced by electric stimulus.

Stimulus by contact or friction induces a retardation which is, relatively speaking, moderate. On the cessation of stimulus the normal rate of growth is restored within an hour.

Intense stimulation caused by the wound gives rise to greater and more persistent retardation of growth.

XVI.—ACTION OF LIGHT ON GROWTH

By

SIR J. C. BOSE,

Assisted by

GURUPRASANNA DAS.

The next subject of inquiry is the *normal* effect of light on growth. I speak of the normal effect because, under certain definite conditions, to be described in a later Paper, the response undergoes a reversal. The Crescograph is so extremely sensitive that it records the effect of even the slightest variation of light. Thus, as I have already mentioned, the opening of the blinds of a moderately-lighted room induces, within a short time, a marked change in the record of the rate of growth. The conditions of the experiment would thus become more precise if the growth-rate in the absence of light is taken as the normal. The specimens are, therefore, kept for several hours in darkness before the experiment. But this should not be carried to the extent of lowering the healthy tone of the plant.

I shall, in the present Paper, determine the characteristic response to light in variation of growth, the latent period of response, the effects of light of increasing intensity and duration, and the effects of the visible and invisible rays of the spectrum.

METHOD OF EXPERIMENT.

The plant was placed in a glass chamber kept in humid condition. The sources of light employed were: (1) an arc-lamp with self-regulating arrangement for securing steadiness of light, and (2) an incandescent electric lamp. Two inclined mirrors were placed close behind the specimen so that it should be acted on by light from all sides.

NORMAL EFFECT OF LIGHT.

Experiment 78.—I shall first give records obtained with *Kysoor* on the action of light. The first series exhibits the normal rate of growth in darkness; in the next the retarding effect of light is seen in the shortening of spacings, as compared with the normal, between successive dots. The light was next cut off and record taken once more after half an hour. Growth is now seen to have recovered its normal rate (Fig. 75). With regard to the after-effect of light I may say in anticipation that there are two different results, which depend on the physiological condition of the tissue. In a tissue whose tonic condition is below par, the after-effect is an acceleration; but with tissues in an optimum condition, the immediate after-effect is a retardation of the rate of growth. This is specially the case when the incident light is of strong intensity and of long duration.

FIG. 75.—Normal effect of light. N, normal; S, retarded rate of growth in response to light; N, recovery on cessation of light.

DETERMINATION OF THE LATENT PERIOD.

There is a general impression that it takes from several minutes to more than an hour for the light to react on the growing organ. This underestimate must have been due to the want of sufficient delicate means of observation. For my recorders indicate in some cases a response within less than 2 seconds of the incidence of light. This was found, for example, in the record of response given by a seedling of *Cucurbita*, to a flash of ultra-violet light. In the majority of cases the response is observed within 15 seconds of the incidence of light.

Experiment 79.—For the determination of the latent period, a record of the effect of arc light of 30 seconds' duration was taken on a moving plate. It will be noticed (Fig. 76) that a retardation of growth was induced

FIG. 76.—Latent period and time-relations of response to light, applied at thick line. Successive dots at intervals of 5 sec.

within 35 seconds of the incidence of light. The incipient contraction induced by light is thus similar to that induced by any other form of stimulus. Growth became restored to the normal value, 5 minutes after the cessation of stimulus.

EFFECT OF INTENSITY OF LIGHT.

Experiment 80.—I next studied the action of light, the intensity of which was increased in arithmetical progression. The intensity of white light given by a half-watt incandescent electric lamp of 200 candle power, placed at a distance of a metre, is taken as the unit. Much feebler light would have been sufficient, but it would have required much longer exposure. The intensity was increased by bringing the lamp nearer the plant; marks were made on a horizontal scale so that the intensity of incident light increased at the successive marks of the scale as 1 : 2 : 3 : and so on. The duration of exposure was same in all cases, namely, 5 minutes. After each experiment suitable periods of rest were allowed for the plant to recover its normal rate of growth. Records in Fig. 77 show increasing retardation induced by stronger intensities of light. Table XVIII gives the result of a different experiment.

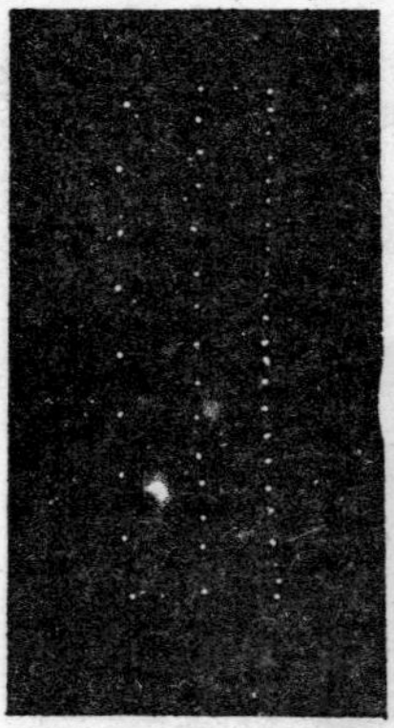

FIG. 77.—Action of light of increasing intensities: 1: 2: 3 in retardation of growth.

TABLE XVIII.—EFFECT OF LIGHT OF INCREASING INTENSITY ON THE RATE OF GROWTH.

Intensity of light.	Rate of growth.
0 (Normal)	0·47 μ per sec.
1 Unit	0·28 μ "
2 "	0·17 μ "
3 "	0·10 μ "
4 "	Arrest of growth.

EFFECT OF CONTINUOUS LIGHT.

Experiment 81.—The continued effect of light of moderate intensity in bringing about increasing retardation of growth will be seen in Fig. 78(*b*) side by side with the record of effect of continuous electric stimulation (Fig. 78*a*) on growth. In both the cases the effect of continuous stimulation is seen to be the same, namely, a growing retardation, which in the given instances culminated in arrest of growth. This is true of stimulus of moderate intensity. Under a more intense stimulation the incipient contraction does not end in a mere arrest of growth, but the responding organ undergoes an actual shortening.

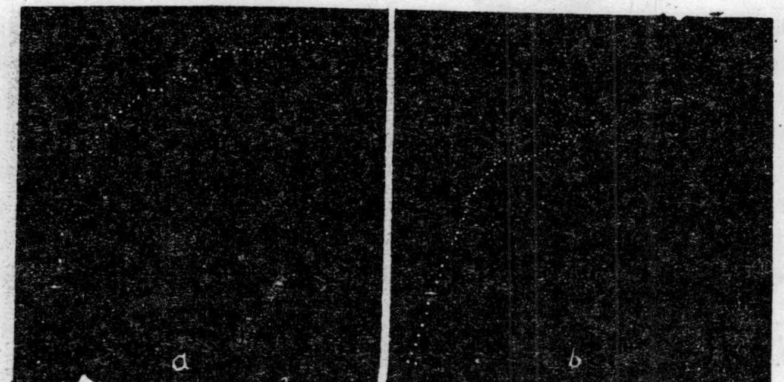

FIG. 78.—Effects of continuous (*a*) electric and (*b*) photic stimulation of moderate intensity, taken on a moving plate.

EFFECTS OF DIFFERENT RAYS OF THE SPECTRUM.

Different observers have found* that it is the more refrangible rays which exercise the greatest influence upon growth and tropic curvature. The relative effects of different lights will, however, become more precise from the curves of response to the action of different rays. For this purpose, I first employed monochromatic lights from different parts of the spectrum, produced by prism of high dispersion. In practice, the usual colour filters were found very convenient, as they allowed the application of more intense light. A thick stratum of bichromate of potash solution

*PFEFFER—Physiology of Plants—Vol. II., p. 104 (English Translation).

transmitted red rays, a thinner stratum allowed the transmission of yellow in addition; ammoniated copper sulphate solution allowed the blue and violet rays to pass through. It should be borne in mind that certain complicating factors are introduced by the incidence of light on the organ; there may be a slight rise of the temperature. We have seen however that moderate rise of temperature induces an acceleration of the rate of growth (p. 175). I shall later describe other experiments which will demonstrate the antagonistic effects of light and warmth on growth. Warmth again may induce a certain amount of dessication, but this is reduced to a minimum by maintaining the plant-chamber in a humid condition. The heating effect of the red is, relatively speaking, much greater than that of the blue rays. But in spite of this it is found that while red rays are practically ineffective, the blue rays are most effective in inducing responsive retardation of growth.

Effect of red and yellow light.—These rays had little or no effect in inducing variation of growth.

Effect of blue light: Experiment 82.—The blue rays exerted a marked retarding effect on growth. Light was applied for 34 seconds and retardation was initiated within 14 seconds of the incidence of light, and the retarded rate was two-fifths of the normal (Fig. 79B).

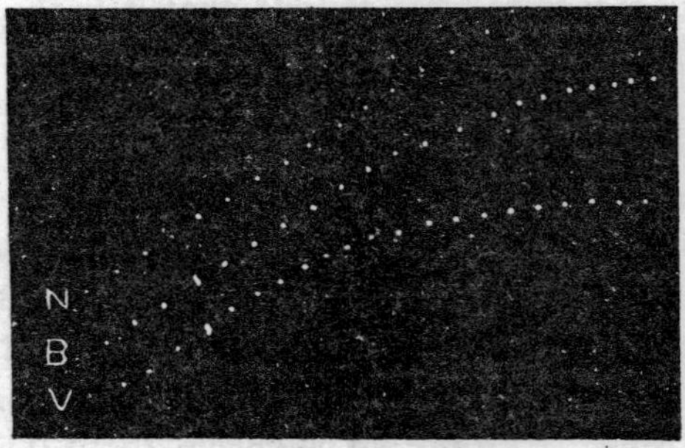

FIG. 79.—N, normal. B, effect of blue light, and V, of ultra-violet light. The records are on a moving plate.

Effect of ultra-violet light : Experiment 83.—Ultra-violet light was obtained from a quartz mercury vapour lamp. The effect of this light in retardation of growth was very marked. Response was induced within 10 seconds, the maximum retardation being one-sixth of the normal rate (Fig. 79V).

Effect of infra-red rays : Experiment 84.—In passing from the most refrangible ultra-violet to the less refrangible red rays, the responsive retardation of growth undergoes a diminution and practical abolition. Proceeding further in the infra-red region of thermal rays, it is found that these latter rays become suddenly effective in inducing retardation of growth.

A curve drawn with the wave length of light as abscissa, and effectiveness of the ray as ordinate shows a fall towards zero as we proceed from the ultra-violet wave towards the red ; the curve, however, shoots up as we proceed further in the region of the infra-red. In connection with this it should be remembered that while the thermal rays induce a retardation of growth, rise of temperature, up to an optimum point, gives rise to the precisely opposite reaction of acceleration of growth.

The relative effectiveness of various rays on growth will be seen more strikingly demonstrated in records of phototropic curvature to be given in a succeeding Paper.

SUMMARY.

The normal effect of light is incipient contraction or retardation of the rate of growth.

The latent period may in some cases be as short as 2 seconds. In large number of cases it is about 15 seconds. The latent period is shortened under stronger intensity of light.

Increasing intensity of light induces increasing retardation and arrest of growth. Under continued action of light of strong intensity the growing organ may undergo an actual shortening.

In these reactions the action of stimulus of light resembles the effects of electric and mechanical stimuli.

The ultra-violet rays induce the most intense reaction in retardation of growth. The less refrangible yellow and red rays are practically ineffective. But the infra-red rays induce a marked retardation of growth.

The effects of light and warmth are antagonistic. The former induces a retardation and the latter an acceleration of growth.

XVII.—EFFECT OF INDIRECT STIMULUS ON GROWTH

By

SIR J. C. BOSE,

Assisted by

GURUPRASANNA DAS.

It has been shown that the direct application of stimulus gives rise in different organs to contraction, diminution of turgor, fall of motile leaf, electro-motive change of galvanometric negativity, and retardation of the rate of growth. I shall now inquire whether Indirect stimulus, that is to say, application of stimulus at some distance from the responding organ, gives rise to an effect different from that of direct application.

MECHANICAL AND ELECTRICAL RESPONSE TO INDIRECT STIMULUS.

I have already described the effect of Indirect stimulus on motile organs (p. 136). A feeble stimulus applied at a distance was found to induce an erectile movement or positive response of the leaf of *Mimosa* or of the leaflet of *Averrhoa*. This reaction is indicative of increase of turgor, an effect which is diametrically opposite to the diminution of turgor induced by the effect of Direct stimulus. It was also shown that an increase in the intensity of Indirect stimulus or a diminution of the intervening distance brought about a diphasic response, positive followed by negative. Direct stimulus gave rise only to a negative response.

Electric response to Indirect stimulus.—I have already explained how an identical reaction finds diverse expression in mechanical and electrical response, or in responsive variation of the rate of growth. It is of interest in this connection to state that my attention was first directed to the characteristic difference between the effects of Direct and Indirect stimulus from the study of electric response of vegetable tissues. I found that while *Direct* stimulus induced negative electric response, *Indirect* stimulus gave rise to a positive response. The clue thus obtained led to the discovery of positive mechanical response under Indirect stimulus.

Experiment 85.—The records given in Fig. 80, exhibit

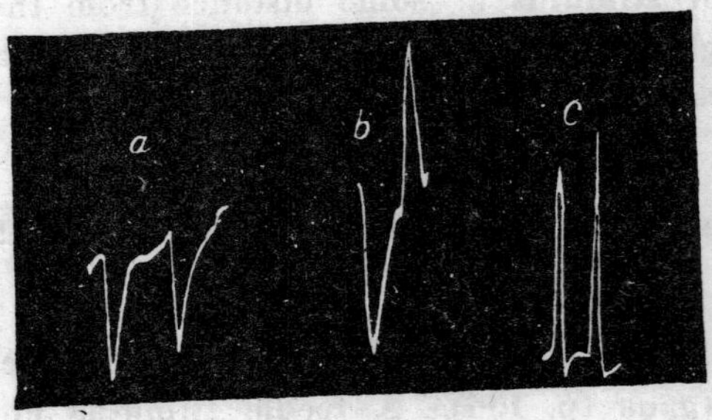

FIG. 80.—Electric response of *Musa* (*a*) Positive, (*b*) diphasic, (*c*) negative.

the electric response given by vegetable tissues. On application of feeble stimulus at a distance from the responding point, the response was by galvanometric positivity. Under stronger stimulus the response became diphasic, positive followed by negative. Direct stimulus induced a negative response.

VARIATION OF GROWTH UNDER INDIRECT STIMULUS.

Since the responsive reactions of growing and non-growing organs are, as we shall find later, fundamentally similar, I expected that Indirect stimulus would give rise in a growing organ to an effect which would be of opposite sign to that induced by Direct stimulus—an acceleration, instead of retardation of growth ; that would correspond to the positive mechanical and electrical responses to Indirect stimulus given by pulvinated organs and by ordinary vegetable tissues. The account of the following typical experiment will show that my anticipations have been fully verified.

Experiment 86.—I took a growing bud of *Crinum* and determined the region of its growth activity ; lower down a region was found where the growth had attained its maximum and may, therefore, be regarded as indifferent region. I applied two electrodes in this indifferent region about 1 cm. below the region of growth. On application of moderate electric stimulus of short duration the response was by an acceleration of growth which persisted for nearly a minute, after which there was a resumption of the normal rate of growth. In this particular case the interval of time between the application of stimulus and the responsive acceleration of growth was 12 seconds. The interval varies in different cases from one second to 20 seconds or more, depending on the intervening distance between the point of application of stimulus and the responding region of growth. I give a record (Fig. 81)

FIG. 81.—Effect of Indirect and Direct stimulus on growth of *Crinum*, taken on a moving plate. Dotted arrow shows application of Indirect stimulus with consequent acceleration of growth. Direct application of stimulus at the second arrow induces contraction and subsequent retardation of rate of growth. Successive dots are intervals of 5″. (Magnification 2,000 times).

obtained in a different experiment which shows in an identical specimen, (1) an acceleration of growth under Indirect and (2) a retardation of growth under Direct stimulus.

TABLE XIX —ACCELERATING EFFECT OF INDIRECT STIMULUS ON GROWTH (*Crinum*).

Specimen.	Condition of experiment.	Rate of growth.
I ...	Normal	0·21 μ per second.
	After Indirect stimulus...	0·26 μ " "
II ...	Normal	0·25 μ " "
	After Indirect stimulus...	0·30 μ " "

It is thus seen that the effect of Indirect stimulus on growth-variation is precisely parallel to that obtained with the response of sensitive plant; that is to say, the effect induced by a feeble stimulus applied at a distance from the growing region is a positive variation or acceleration of growth. The effect becomes converted into negative or retardation of growth when the stimulus is Direct, *i.e.*, when applied to the responding region of growth; under intermediate conditions, the growth-variation I find to be diphasic, a positive acceleration followed by a negative retardation. This is found true not merely in the case of a particular form of stimulus but of stimuli as different as mechanical, thermal, electric, and photic.

I shall in a subsequent paper formulate a generalised Law of Effects of Direct and Indirect Stimulus. From the experiments already described it is seen that:

Direct stimulus induces negative variation of *turgor, contraction, fall of leaf of* Mimosa, *electric change of galvanometric negativity, and retardation of the rate of growth.*

Indirect stimulus induces positive variation of turgor, expansion, erection of leaf of Mimosa, *electrical change of galvanometric positivity, and acceleration of the rate of growth.*

It is seen that Indirect stimulus gives rise to dual reactions, seen in positive and negative responses; of these the negative is the more intense. When the intervening distance is reduced, the resulting response becomes negative; this is due not to the absence of the positive, but to its being masked by the predominant negative. From the principle of continuity, this will also hold good in the limiting case, where by the reduction of the intervening distance to zero, the stimulus becomes Direct. In other words, Direct stimulus should also give rise to both positive and negative reactions. Of these the positive is masked by the predominant negative.

So much for theory; is it possible to unmask the contained positive in the resulting negative response under Direct stimulus? This important aspect of the subject will be dealt with in the following Paper.

SUMMARY.

The application of Direct stimulus gives rise to an electric response of galvanometric negativity. The application of stimulus at a distance from the responding point, *i.e.*, Indirect stimulus, gives rise to positive electric response.

The mechanical responses of sensitive plants also exhibit similar effects, *i.e.*, a negative response under Direct, and positive response under Indirect stimulus.

In the responsive variation of growth, Direct stimulus induces a retardation, and Indirect stimulus an acceleration of the rate of growth.

The effects of Direct and Indirect stimulus on vegetable organs in general are as follows:

Direct stimulus induces negative variation of turgor, contraction, fall of leaf of *Mimosa*, electric change of galvanometric negativity, and retardation of the rate of growth.

Indirect stimulus induces positive variation of turgor, expansion, erection of leaf of *Mimosa*, electrical change of galvanometric positivity and acceleration of the rate of growth.

XVIII.—RESPONSE OF GROWING ORGANS IN STATE OF SUB-TONICITY

By

SIR J. C. BOSE.

The normal response of a growing organ to Direct stimulus is *negative*, that is to say, a retardation of the rate of growth. This is the case under forms of stimuli as diverse as those of mechanical and electric shocks, and of the stimulus of light.

ABNORMAL ACCELERATION OF GROWTH UNDER STIMULUS.

After my investigations on the normal retarding effect of light on growth, I was considerably surprised to find the responses occasionally becoming *positive*, an acceleration instead of retardation of growth. I shall first give accounts of such positive responses and then explain the cause of the abnormality.

Abnormal acceleration under stimulus of light: Experiment 87.—A rather weak specimen of *Kysoor* was exposed to the action of light of 5 minutes' duration. This induced an abnormal acceleration in the rate of growth from 0·30 μ to 0·40 μ per second. But continuous exposure to light for half an hour brought about the normal effect of retardation. In trying to account for this abnormality in response I found that while specimens of *Kysoor* in a vigorous state of growth of about 0·8 μ per second exhibit normal retardation of growth under light, the particular specimen which exhibited the abnormal positive response had a much feebler rate of growth of 0·30 μ per second. As activity of growth in a plant is an index of its healthy tone, a feeble rate of growth must be indicative of tonicity below par. The fact that plants in sub-tonic condition exhibit abnormal acceleration of growth under stimulus will be seen further demonstrated in the next experiment.

In the parallel phenomenon of the response of pulvinated organs we found that under condition of sub-tonicity, the response becomes positive and that this abnormal positive is converted into normal negative in consequence of repeated stimulation. In growth, response likewise the abnormal acceleration of growth under light in the sub-tonic specimen of *Kysoor* was converted into normal retardation after continuous stimulation for half an hour. From the facts given above, we are justified in drawing the following conclusions:

(1) That while light induces a *retardation* of growth in a tissue whose tonic condition is normal or above par, it brings about an *acceleration* in a tissue whose condition is below par.

(2) That by the action of the stimulus of light itself a sub-tonic tissue is raised to a condition at par, with the concomitant restoration of normal mode of response by retardation of growth.

Another important question arises in this connection: Is the restoration of normal response due to light as a form of stimulus, or to its photo-synthetic action? An answer to this is to be found from the results of an inquiry, whether a very different form of stimulus which exerts no photo-synthetic action, such as tentanising electric shocks, also induces a similar acceleration of growth in a sub-tonic tissue.

The normal retarding effect of electric stimulus on specimens in active state of growth was demonstrated in record given in Fig. 72, where the normal rate was found greatly reduced after stimulation.

Abnormal acceleration of growth under electric stimulus; Experiment 88.—For my present purpose I took a sub-tonic specimen of seedling of wheat, its rate of growth being as low as 0·05 μ per second. After electric stimulation the rate was found enhanced to 0·12 μ per second, or about two and-a-half times. I give (Fig. 82) two records

FIG. 82.—Enhancement of rate of growth in sub-tonic specimens of wheat seedling. First series of record on stationary, second series (*b*) on moving plate N, record before stimulation. S, after stimulation.

obtained with two different specimens. In the first, the record was taken on a stationary plate (Fig. 82); the closeness of successive dots in N show the feeble rate of growth of the sub-tonic specimen, the wider spacing after stimulation, S, exhibit the induced enhancement of growth.

In the second experiment the records (Fig. 82*b*) were taken on a moving plate. The specimen was so extremely sub-tonic, that its normal record N appears almost horizontal. The greater erection of the curve, S, after stimulation demonstrates the induced acceleration of growth.

TABLE XX.—ACCELERATION OF GROWTH BY STIMULUS IN SUB-TONIC SPECIMENS.

Specimen.	Stimulus.	Rate of growth.
Wheat seedling	Normal	0·05 μ per sec.
	After electric stimulation	0·12 μ " "
Kysoor ...	Normal	0·30 μ per sec.
	After 5′ exposure to light	0·40 μ " "
	" 30′ " "	0·27 μ " "

In my previous Paper on the 'Modifying Influence of Tonic Condition' I showed that while the response of the primary pulvinus of *Mimosa* in normal condition is *negative*, i.e., by contraction, diminution of turgor, and fall of the leaf, the response of a sub-tonic specimen is *positive*, that is to say, by expansion, enhancement of turgor, and erection of the leaf. I have shown further that in a sub-tonic specimen the action of stimulus itself raises the tissue from below par to normal or even above par, with the conversion of abnormal positive to normal negative response.

I have in the present Paper shown that a parallel series of reactions is seen in the response of growing organs. In vigorously growing specimens the action of stimulus is *negative*, i.e., incipient contraction, diminution of turgor, and retardation of the rate of growth. But in sub-tonic specimens, with enfeebled rate of growth, the effect of stimulus is *positive*, i.e., by expansion, enhancement of turgor, and acceleration of the rate of growth. Continuous stimulation also raises the sub-tonic growing tissue to a condition at par, converting the response from abnormal positive to normal negative.

It was also explained that every stimulus gave rise to dual reactions, positive and negative, and that in a highly excitable tissue the positive is masked by the predominant negative. The positive, or A-effect, is generally described as a "building up" process. By choosing a sub-tonic specimen, I have been able to unmask the positive, A. In the case of sub-tonic growing organs the positive, A, is literally a building up process, giving rise to an acceleration of growth.

From these facts and others given previously it will be seen that the abnormal response of acceleration of growth under stimulus is by no means accidental or fortuitous but is a definite expression of an universal reaction, characteristically exhibited by all tissues in a condition of sub-tonicity.

CONTINUITY BETWEEN ABNORMAL AND NORMAL RESPONSES.

A given plant-tissue may exist in widely different conditions of tonicity. Let us take two extreme conditions, the optimum and the minimum. The tonic level will be at its lowest at the minimum, where growth will be at a standstill. The range between the optimum and minimum will be very extended; hence strong and long continued stimulation will be necessary to raise the tissue from the tonic minimum to the optimum level. There are innumerable grades of tonicity between the optimum and minimum. Within this wide range the characteristic response will be the abnormal positive. As we approach the optimum, the range for positive response will become circumscribed, and the intensity and duration of stimulus necessary to convert the positive to negative will be feebler and shorter. It will be very seldom that a plant is likely to be found at the optimum. Hence plants in general may be expected to give a feeble positive response under sub-minimal stimulus.

These considerations led me to look for the positive response under sub-minimal stimulation; the tracings which I have obtained with my highly sensitive Crescograph and other recorders show that my anticipations have been justified.

Positive response under sub-minimal stimulus: Experiment 89.—In normal specimens, light of strong intensity induces a retardation of growth. When the source of light is placed at a distance, the intensity of light undergoes great diminution. Under the action of such feeble stimulus I obtained an acceleration of growth even in specimens which may be regarded as moderately vigorous (Fig. 83). Similar acceleration of growth was also

FIG. 83—Acceleration of growth under sub-minimal light stimulus. Record on moving plate; stimulus applied at 5th dot, and subsequent erection of curve exhibits acceleration of growth. Last part of curve shows recovery of normal growth on cessation of stimulus.

obtained under feeble electric stimulation. The response is reversed to normal negative by increasing the intensity or duration of stimulus. Very feeble stimulus thus induces an acceleration and strong stimulus a retardation of growth. I have frequently obtained positive mechanical and electrical responses under sub-minimal stimulation. As chemical substances often act as stimulating agents, the opposite effects of the same drug in small and large doses may perhaps prove to be a parallel phenomenon.

It has been shown that stimulus induces simultaneously both A- and D-effects, with the attendant positive and negative responsive reactions, alike in pulvinated and in growing organs. A tissue, in an optimum condition, exhibits only the resultant negative response ; the comparatively feeble positive is imperceptible, being masked by the predominant negative ; but with the decline of its tone excitability diminishes, with it the D-effect, and we get the A-effect unmasked, resulting response then becomes diphasic. In extreme sub-tonic condition, it exhibits only the positive. The sequence is reversed when we begin with a tissue in a state of extreme sub-tonicity, which first exhibits only the positive. Successive stimulations continually exalt the tonic condition, the subsequent responses becoming, diphasic, and, with the attainment of optimum tone, a resultant negative response, As a further verification of the simultaneous existence of both A-and D-effects, it has been shown that in ordinary tonic condition a *sub-minimal* stimulus gives rise only to positive response ; this becomes converted into normal negative under moderate stimulation.

I have described the action of stimulus on tissues in which, on account of sub-tonicity, growth has become enfeebled. I shall next take up the question of effect of stimulus on tissues in which growth, on account of extreme sub-tonicity, has been brought to a state of standstill.

SUMMARY.

The modifying influence of tonic condition on response is similar in pulvinated and growing organs.

The motile organ of *Mimosa* in a condition of sub-tonicity, exhibits a *positive* response, by expansion, increase of turgor, and erection of the leaf. Continuous stimulation converts the abnormal *positive* to normal *negative.*

In sub-tonic growing organs stimulus likewise induces a *positive* response, by expansion, increase of turgor and acceleration of the rate of growth. Continuous stimulation converts the abnormal acceleration to normal retardation.

Sub-minimal stimulus tends to induce even in normal tissues, an acceleration of rate of growth. Stimulus of moderate intensity induces in the same tissue the normal retardation of growth.

XIX.—RESUMPTION OF AUTONOMOUS PULSATION AND OF GROWTH UNDER STIMULUS

By

SIR J. C. BOSE.

The autonomous activity of growth is ultimately derived from energy supplied by the environment. The internal activity may fall below par with consequent diminution or even arrest of growth; this condition of the tissue I have designated as sub-tonic. The inert plant can only be stirred up to a state of activity by stimulus from outside; and we saw that under the action of stimulus the rate of growth of a sub-tonic tissue was enhanced.

As the general question of depression of autonomous activity and its restoration by the action of stimulus is of much theoretical importance, I shall describe experiments carried out on a different form of autonomous activity, seen in spontaneous pulsation of the lateral leaflets of *Desmodium gyrans*. Under favourable conditions of light and warmth these leaflets execute vigorous movements, the period of a single pulse varying from one to two minutes. As the energy for this activity is ultimately derived from the environment, it is clear that isolation from the action of favourable environment will bring about a gradual depletion of energy with concomitant decline and ultimate cessation of spontaneous movement. For this we may keep the plant in semi-darkness; we may further hasten the rundown process by isolating the leaflet from the parent plant. A leaflet immersed in water was kept in a dimly lighted room; it was attached by a cocoon thread to the recording lever of an Oscillating Recorder to be fully described in the next Paper. The pulsation continued even in this isolated condition for about 48 hours, after which the spontaneous movement came to a stop. Further experiments showed that the arrest of pulsation was not indicative of mortality but of 'latent life' in a state of suspense, to be stirred up again by shock stimulus into throbbing activity.

REVIVAL OF AUTONOMOUS PULSATION UNDER STIMULUS.

Experiment 90.—In figure 84, is a seen record of the

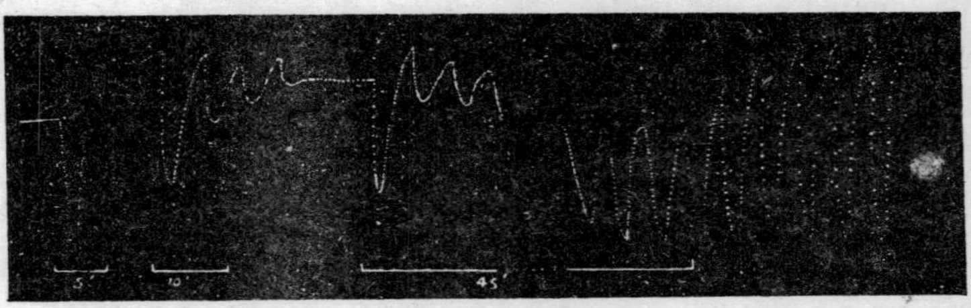

FIG. 84.—Renewal of autonomous activity in *Desmodium gyrans* at standstill by action of light. Up-curve represents up-movement. The horizontal lines below represent durations of exposure to light.

action of light on the sub-tonic *Desmodium* leaflet at standstill. A narrow pencil of light from electric arc was first thrown on the lamina in which the presence of chlorophyll rendered photo-synthetic action possible. This had no effect on the renewal of pulsation. But the autonomous activity was revived by the action of light on the pulvinule. This preferential effect on pulvinule showed that the renewal of activity was due not to photo-synthesis but to the stimulating action of light. The pulsation was also restored by chemical stimulants, such as dilute ether, and solution of ammonium carbonate.

As regards the action of light, the pulsation continued for a time, even on the cessation of light. This persistence of autonomous activity increases with the intensity and duration of incident stimulus, that is to say with the amount of incident energy. In the present case a duration of five minutes' exposure gave rise to a single pulsation, after which the movement of the leaflet came to a stop. The next application lasted for ten minutes and this gave rise to four pulsations, two during application, and two after cessation of light. The next application was for forty-five minutes, and the pulsation persisted for nearly an hour after the cessation of light. The experiments on sub-tonic specimens show clearly that the energy supplied by the environment becomes as it were latent in the plant, increasing its potentiality for work.

The renewal of autonomous activity in a sub-tonic tissue by the action of external stimulus, will be found in every way parallel to the renewal of growth in a sub-tonic organ.

REVIVAL OF GROWTH UNDER STIMULUS.

Renewal of growth under stimulus: Experiment 91.—I find that application of electric stimulus renews growth in specimens where, on account of extreme sub-tonicity growth has come to a state of standstill. The resumption of growth in grass haulms under the stimulus of gravity is a phenomenon probably connected with the above. The causes which bring about cessation of growth in a mature organ are unknown; that there is a potentiality of growth even in a fully grown grass haulm is evidenced by the fact of its renewed growth under fresh stimulation. That this is not an exceptional phenomenon appears from the record which I obtained with a fully grown style of

Datura alba. I subjected it to periodic stimulation, and obtained from it a series of contractile responses. After recovery from stimulus it regained its normal length which remained constant for some time as seen in the horizontal base-line. But as a result of successive stimulations, the mature style resumed its growth with increasing acceleration. This is seen in the recovery overshooting its former horizontal limit (Fig. 85).

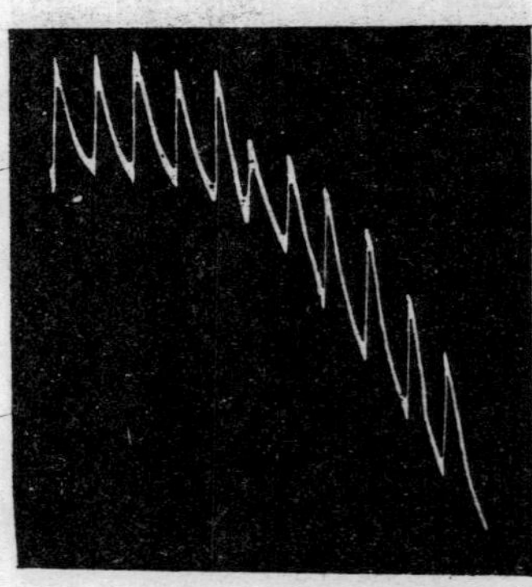

FIG. 85.—Record of responses of a mature style in which growth had come to a stop. Up-curve shows contraction under stimulus. Renewal of growth at sixth response, after which growth-elongation is shown by the trend of the base-line downwards.

From the investigations that have been described in this and in the previous Papers an insight is obtained into the complexity of response arising from various factors. It has been shown that the sign of response is modified by the intensity of stimulus, by its point of application, and by the tonic condition of the responding tissue. The fundamental reactions have been found to be essentially the same in pulvinated, in growing and non-growing organs. The results described enable us to enunciate general Laws of Effects of Direct and Indirect stimulus on tissues in normal and in sub-tonic condition.

LAWS OF EFFECTS OF DIRECT AND INDIRECT STIMULUS.

1. THE EFFECT OF DIRECT STIMULUS IS NEGATIVE VARIATION OF TURGOR, NEGATIVE MECHANICAL AND ELECTRICAL RESPONSE, NEGATIVE VARIATION, OR RETARDATION OF RATE OF GROWTH.

a. SUB-MINIMAL STIMULUS GIVES POSITIVE RESPONSE.

b. POSITIVE RESPONSE IS ALSO GIVEN BY A TISSUE IN A SUB-TONIC CONDITION: CONTINUOUS STIMULATION CONVERTS THE ABNORMAL POSITIVE TO NORMAL NEGATIVE RESPONSE.

c. AUTONOMOUS ACTIVITY IN A STATE OF STANDSTILL, MAY BE REVIVED BY STIMULUS.

d. THE EFFECTS OF STIMULUS AND WARMTH ARE ANTAGONISTIC.

2. THE EFFECT OF INDIRECT STIMULUS IS POSITIVE VARIATION OF TURGOR, POSITIVE MECHANICAL AND ELECTRICAL RESPONSE AND POSITIVE VARIATION OR ACCELERATION OF RATE OF GROWTH.

I have referred to the fact previously demonstrated, that while Direct stimulus induces contraction and retardation of growth, moderate rise of temperature induces the opposite effect of expansion and acceleration of growth. Further demonstration of the antagonistic effects of stimulus and warmth will be given in the next Paper.

SUMMARY.

The autonomous activity of pulsating leaflet of *Desmodium gyrans* comes to a stop under depletion of internal energy. A cut leaf isolated from the plant maintains the rhythmic activity of its leaflets for about 48 hours, after which there is an arrest of movement.

In this state of sub-tonicity the arrested autonomous activity is revived under the action of various stimuli. Thus the incidence of light on the pulvinule initiates pulsatory movements, which persists for a time even on the cessation of stimulus. This persistence of autonomous activity increases with the intensity and duration of stimulus to which the leaflet had been subjected.

The arrested autonomous activity of growth may often be revived by the action of stimulus. Thus the arrested growth in a mature style or *Datura alba* was renewed by electric stimulation.

XX.--ACTION OF LIGHT AND WARMTH ON AUTONOMOUS ACTIVITY

By

SIR J. C. BOSE.

In the preceding Paper I have shown the essential similarity of effect of stimulus on autonomous activity of the *Desmodium* leaflet, and of the growing organ. It was shown how stimulus revived the pulsatory activity of *Desmodium* leaflet in a state of standstill, in the same way as it renewed the arrested growth-activity.

THE OSCILLATING RECORDER.

The investigation of this subject was rendered possible by the successful device of my Oscillating Recorder. A very light glass fibre was used for the construction of the lever, which was supported on jewel bearings. The short arm of the lever was 2 cm. in length, and the long arm 8 cm. This gave a magnification of 4 times. But it is quite easy to increase the magnification to 10 times or more.

The pull exerted by the pulsating leaflet is extremely slight, and the relatively heavy lever made of steel wire used in the Resonant Recorder is not well-suited for our purpose. The pulsation of the leaflet is relatively slow, being once in two minutes or so. The intermittent contact of ten times in a second, given by the Resonant Recorder, is therefore too quick. In the Oscillating Recorder the intermittence was, therefore, reduced to once in a second, or once in five seconds, the recording plate itself being made to move to-and-fro at this rate. The carrier of the plate-holder slides backwards and forwards on ball bearings ; a wheel in the clockwork connected with an eccentric is released periodically, at intervals which may be varied between one and five seconds. By the action of the eccentric, the plate carrier approaches the writing lever with diminishing speed till the movement is zero at the contact. This contrivance is essential, since any sudden shock of the plate against the lever is apt to give rise to after-vibrations of the writer. The plate carrier is quickly withdrawn after the production of a dot on the smoked glass plate by contact with the writing lever.

The clockwork is governed by a revolving fan which can be gradually opened out by a regulating screw. The speed can thus be adjusted within wide limits, and maintained constant and at any desired speed. A second set of wheels connected with the clockwork moves the plate-holder in a lateral direction. A series of records may thus be taken for fifteen minutes, half an hour, or an hour.

The record obtained in this way is very perfect. Not only is the effect of an external agent shown by variation in the amplitude and frequency of pulsations, but the change of speed in any phase of the pulse becomes automatically recorded.

RECORD OF PULSATION OF *DESMODIUM GYRANS.*

The whole plant can not be conveniently manipulated for different investigations. It is, however, possible using the precautions described below to use the detached petiole carrying the pulsating leaflets. The terminal large leaf may also be removed. The necessary amputation is often followed by an arrest of pulsation. But as in the case of isolated heart in a state of standstill, the movement of the leaflet may be revived by the application of internal hydrostatic pressure. Under these conditions, the rhythmic pulsations may easily be maintained uniform for many hours.

The petiole carrying the leaflet is mounted water-tight in the short arm of an U-tube filled with water; for producing internal hydrostatic pressure in the plant the height of water in the longer arm is suitably raised. The U-tube holding the specimen may be adjusted up and down, and laterally. A hinged support also allows the specimen to be placed at any inclination. The movement of the leaflet, it is to be remembered, does not always take place in a vertical direction. The object of the mechanical adjustments is to place the specimen at such an angle that its up and down movements when in a straight line should be vertical, or have its long axis vertical when the movement is elliptical. It is important that the specimen should be illuminated equally from all sides; for one-sided illumination causes a bending over of the leaflet towards light.

The pulvinule of the leaflet acts like the pulvinus of *Mimosa*, that is to say, the leaflet undergoes a sudden fall to down position by the contraction of the more effective lower half of the pulvinule; the 'up' position denotes recovery and expansion of the more effective half. The up-and-down movements of the leaflet correspond to the diastolic and systolic movements of the animal heart. There is, indeed, as I have shown elsewhere* a very close resemblance between the activities of rhythmic tissue in the plant and in the animal.

EFFECT OF DIFFUSE LIGHT ON PULSATION OF *DESMODIUM*.

Experiment. 92.—For the study of effect of light on *Desmodium,* I first obtained record in darkness. A horizontal beam of divergent light from an arc lamp placed at a distance of 200 cm. was made to act diffusely on the leaf from all sides. This was done by means of three inclined mirrors, the first throwing the light vertically downwards, the

* Bose—Irritability of Plants—p. 295.

second vertically upwards, and the third horizontally forward from the side opposite the lantern. The effect of light is seen demonstrated in Fig. 86.

FIG. 86.—Effect of light on pulsation of *Desmodium* leaflet. Duration of application of light is represented by the horizontal line. Up-curve represents diastolic expansion and down-curve systolic contraction. Note contractile effect of light in diminution of amplitude and reduction of diastolic limit of pulsation.

Light was applied at the second pulsation. It will be seen that light retards or arrests the autonomous activity. On the cessation of light the normal activity was found to be gradually restored. It is of much interest to note here the similarity of action of light on autonomous activity of the leaflet of *Desmodium* and of a growing organ. In both, we find that while in the sub-tonic condition of the tissue the effect of light is to enhance or renew the autonomous activity of growth and pulsation, in normal condition the effect is to retard it.

Inspection of the record exhibits another very interesting characteristic. We saw that light retarded growth by inducing an incipient contraction. In the *Desmodium* leaflet the contractile reaction of light is exhibited by the characteristic modification of its pulsations. The duration of application of light is represented by the horizontal line. In Fig. 86 the up-curve represents up-movement of diastolic expansion, and the down-curve of systolic contraction. The contractile reaction of light is seen to counteract the normal expansion, with diminution of diastolic limit of pulsation.

EFFECT OF RISE OF TEMPERATURE ON PULSATION.

It has been shown that while rise of temperature up to an optimum enhanced the rate of growth, the effect of light was to retard it. Hence the effects of light and warmth are antagonistic.

Effect of rise of temperature on pulse-record : Experiment 93.—In studying the effect of rise of temperature on the pulsation of leaflets of *Desmodium*, we discover similar

FIG. 87.—Effect of rise of temperature on pulsation of leaflet of *Desmodium gyrans*. Horizontal line represents the duration of gradual rise of temperature from 30°C. to 35°C. Note the expansive effect of rise of temperature in reduction of systolic limit of pulsation.

antagonistic reactions of light and warmth. The leaflet was placed in a plant-chamber with an electric arrangement for gradual rise of temperature. The first two records were taken in the normal temperature of the room, which was 30°C. The temperature was gradually raised to 35°C, the record being taken all the time. It will be seen (Fig. 87) that the effect of warmth is diametrically opposite to that of light. The record in Fig. 86 exhibited the *contractile* effect of light by reducing the diastolic limit of expansion. In the present case the *expansive* reaction of warmth is exhibited by the reduction of systolic limit of contraction. The temperature of the plant chamber was now allowed to return to 30°C., and we observe the gradual restoration of normal systolic limit of contraction.

SUMMARY.

Two different effects are found in the action of the stimulus of light alike on the autonomous activity of leaflet of *Desmodium gyrans* and of growing organs. In condition of sub-tonicity light renews pulsation of *Desmodium* and enhances the activity of growth. In normal tonic condition the effect induced is the very opposite, light causing an arrest of pulsation and retardation or arrest of growth.

The contractile effect of light is seen not only in the retardation of growth, but also by the characteristic modification of pulsation of *Desmodium* in the diminution of diastolic limit of expansion.

The antagonistic reactions of light and warmth are found not only in growth but also in the rhythmic activity of *Desmodium gyrans*. In the pulsation of *Desmodium* the contractile effect of light induces a rapid diminution of the diastolic limit of expansion, while the expansive reaction of warmth brings about a marked reduction of the systolic limit in successive pulsations.

XXI.—A COMPARISON OF RESPONSES IN GROWING AND NON-GROWING ORGANS

By

SIR J. C. BOSE,

Assisted by

GURUPRASANNA DAS.

I have in the preceding series of Papers demonstrated the effects of various forms of stimuli on growth. I have also given accounts of numerous reactions which are extraordinarily similar, in growing and non-growing organs. In fact certain characteristic reactions observed in motile pulvinus of *Mimosa* and other 'sensitive' plants led to the discovery of the corresponding phenomena in growing organs. For fully realising the essential similarity of responses given by all plant-organs, growing and non-growing, I shall give here a short review of the striking character of the parallelism.

1. The incipient contraction of a growing organ under stimulus culminates in a marked shortening of the organ.

2. The similarity of contractile responses in growing and pulvinated organs.

3. Similar modification of both under condition of sub-tonicity.

4. The opposite effects of Direct and Indirect stimulus, both in motile and in growing organs.

5. The exhibition by all plant-organs of *negative* electric response under Direct, and *positive* electric response under Indirect stimulus.

6. Similar modification of autonomous activity in *Desmodium gyrans* and in growing organs under parallel conditions.

7. Similar excitatory effects of various stimuli on pulvinated and growing organs.

8. Similar discriminative effects of different rays of light in excitation of motile and growing organs.

CONTRACTILE RESPONSE OF GROWING AND NON-GROWING ORGANS.

I have shown (page 198) that a growing organ under stimulus, undergoes an incipient contraction as shown in the responsive retardation of its rate of growth; that this retardation increases with the intensity of the incident stimulus till growth becomes arrested. Above this critical intensity the induced contraction causes an actual shortening of the organ. There is no breach of continuity in the increasing contractile reaction, which at various stages appears as a retardation, an arrest of growth or a marked shortening of length of the organ.

CONTRACTILE RESPONSE OF PULVINATED AND GROWING ORGANS.

Experiment 94.—In order to show the striking similarity between the response of 'sensitive' *Mimosa* and that of a growing organ, I give a record (Fig. 88) obtained with a

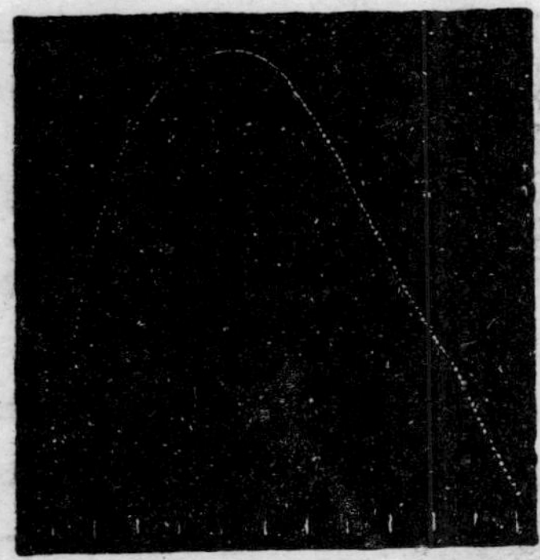

FIG. 88.—Contractile response of growing organ under electric shock. Successive dots at intervals of 4″. Vertical lines below represent intervals of one minute. (Magnification 1,000 times.)

growing bud of *Crinum* under the stimulus of electric shock above the critical intensity. The recorder gave a magnification of a thousand times. In Fig. 88, the normal growth elongation is represented as a down-curve. On the application of stimulus the normal expansion was suddenly reversed to excitatory contraction, the latent period of reaction was one second and the period of the attainment of maximum contraction (apex-time) was 4 minutes. The organ recovered its original length after a further period of seven minutes and then continued its natural growth elongation. Repetition of stimuli gave rise to successive contractile responses which are in every way similar to the mechanical responses of *Mimosa pudica*. The essential similarity of response of pulvinated and growing organs will be seen in the following tabular statement :

TABLE XXI.—TIME RELATIONS OF MECHANICAL RESPONSE OF PULVINATED AND GROWING ORGANS.

Specimen.	Latent period	Apex-time.	Period of recovery.
Motile pulvinus of *Mimosa pudica.*	0·1 sec.	3 secs.	16 minutes.
Motile pulvinus of *Neptunia oleracea.*	0·6 "	180 "	60 "
Growing bud of *Crinum* ...	1·0 "	240 "	7 "

The contraction in growing organs under stimulus is sometimes considerable. Thus in the filamentous corona of *Passiflora quadrangularis* the contraction may be as much as 15 per cent. of the original length. This is not very different from the excitatory reaction of the typically sensitive stamens of the *Cynereæ*, which exhibits a contraction from 8 to 22 per cent.

MODIFICATION OF RESPONSE BY CONDITION OF SUB-TONICITY.

In *Mimosa* the normal response to direct stimulus is *negative*, the leaf undergoing a fall. But sub-tonic specimens exhibit a *positive* response with erection of the leaf. The action of the stimulus itself improves the tonic condition, and the abnormal positive is thus converted into normal negative, through diphasic response (p. 147). Similarly in growing organs, while the normal effect of stimulus is incipient contraction and retardation of growth under condition of sub-tonicity the response is by acceleration of growth. Continuous stimulation converts this abnormal acceleration into normal retardation of growth (p. 225).

EFFECTS OF DIRECT AND INDIRECT STIMULUS.

Direct stimulus induces in *Mimosa* and other 'sensitive' plants a *negative* response. There is a diminution of turgor and contraction in the motile organ, resulting in the fall of leaf. Indirect stimulus, on the other hand, gives rise to a *positive* or erectile response, indicative of increase of turgor and expansion (p. 138).

In growing organs Direct stimulus induces an incipient contraction and retardation of rate of growth; the effect of Indirect stimulus is expansion and accelaration of the rate of growth (p. 216).

The opposite reactions to Direct and Indirect stimulus are also found in the electric response given by all plant organs. Thus while Direct stimulus induces an electromotive change of galvanometric negativity, Indirect stimulus induces the opposite change of galvanometric positivity (p. 214).

MODIFICATION OF AUTONOMOUS ACTIVITY.

The autonomous activity of *Desmodium gyrans* exhibited by the pulsation of its leaflets come to a stop under condition of sub-tonicity. The arrested movement is, however, revived by the action of stimulus (p. 228). The depressed or arrested growth of a growing organ is similarly accelerated or revived by the action of stimulus (p. 230).

In vigorous specimens stimulus induces the opposite effect by retarding or arresting the pulsatory activity or growth.

Warmth induces an effect which is antagonistic to that of stimulus. The contractile effect of stimulus is seen in the pulsations of leaflet *Desmodium* by the reduction of their expansive or diastolic limit, and in growing organs by the retardation of the rate of growth. The expansive effect of warmth is seen in reduction of the systolic limit of *Desmodium* pulsation, and in the acceleration of rate of growth in growing organs (p. 237).

EXCITATORY EFFECTS OF VARIOUS STIMULI ON PULVINATED AND GROWING ORGANS.

Certain agents induce excitation in living tissues, the excitatory change being detected by contraction, or by electromotive variation, or by change of electric resistance, and in growing organs, by the retardation of the rate of growth. In general, the various stimuli which excite animal tissues also excite vegetable tissues.

It has been shown that *every form of stimuli, however diverse, also induces incipient contraction and retardation of the rate of growth.* Thus mechanical irritation, such as friction or wound, induces a retardation of growth (p. 202); they also induce an excitatory contraction in *Mimosa,* attended by the fall of the leaf. Different modes of electric stimulation act similarly on both growing and pulvinated organs. The action of light visible and invisible will presently be seen to react on both alike. And in this connection nothing could be more significant than the discriminative manner in which both the pulvinated and the growing organs respond to certain lights and not to others.

In contrast to the contractile effect of stimulus, certain agents induce the antagonistic reaction of expansion. It has been shown that while stimulus induces a retardation, rise of temperature up to an optimum point, induces an acceleration of the rate of growth. I have also referred to the fact that while the autonomous pulsations of *Desmodium* leaflet exhibit under stimulus a diminution of the extent of the diastolic expansion, warmth on the other hand, induces the opposite effect by diminishing the systolic contraction.

EFFECT OF LIGHT ON PULVINATED ORGANS.

I have referred to the well-known fact that it is the more refrangible portions of the spectrum that are more effective in inducing excitatory reactions and have already given records of the responsive reactions of various lights on growing organs. I shall now give records of the effect of various lights on the pulvinus of *Mimosa pudica.* The amplitude and time relations of the curves of response will give a more precise idea of the quantitative effects of various lights in inducing excitation.

Action of white light : *Experiment 95.*—The source of light was an arc lamp; a pencil of parallel light is made to pass through a trough of alum solution. This process of excluding thermal rays is adopted for the visible rays of the spectrum. Colour filters were also used for obtaining red, yellow and blue lights. The pencil of light is thrown upwards by an inclined mirror on the lower half of the pulvinus. The response is taken by an Oscillating recorder, giving successive dots at intervals of 10 seconds, the magnification employed being 100 times. The pulvinus being subjected to light for 10 seconds gave response by a fall of the leaf (Fig. 89). The response to light

FIG 89.—Effect of white light on the pulvinus of *Mimosa.* Successive dots in this and in the following records are at intervals of 10". (Magnification 100 times)

is thus found to be essentially similar to that induced by electric stimulus, the only difference being in the relative sluggishness of the reply. Electric shock passes instantaneously through the mass of the pulvinus, stirring up the active tissues to responsive contraction. The latent period is, therefore, as short as 0·1 second and the maximum contraction is effected in about 3 seconds. In the case of the stimulus of light the shock-effect is not so great; excitation, moreover, has to pass slowly from the surface of the pulvinus inwards. Hence the latent period is twelve seconds, and the period of maximum contraction is as long as 90 seconds. As the stimulation is moderate, the recovery is effected in 11 minutes, instead of 16 minutes, which is the usual period for *Mimosa* to recover from an electric shock. The important conclusion to be derived from this experiment is, that light is a mode of stimulation end that it induces a responsive contraction, similar to that caused by other forms of stimuli. This contractile response under light is exhibited not merely by the motile pulvinus of *Mimosa*, but by other pulvini as well, such as those of *Erythrina indica*, and of the ordinary bean plant.

Action of red and yellow lights.—The pulvinus gave little or practically no response to these lights.

Action of blue light: *Experiment 96.*—Light was applied for 10 seconds and the amplitude of response was similar to that induced by white light (Fig. 90).

FIG. 90.—Effect of blue light on pulvinus of *Mimosa*.

Action of Ultra-violet rays: *Experiment 97.*—The source of light was a quartz mercury-vapour lamp. The effect was so intense that, to keep the record within the plate, I had to reduce the period of exposure to half, *i.e.*, to five seconds. The responsive movement was initiated within six seconds of the application of light. The intensity and the rapidity of reaction is independently evidenced by the more erect curve of response (Fig. 91).

FIG. 91.—Effect of ultra-violet rays on the pulvinus of *Mimosa*

Action of Infra-red rays: *Experiment 98.*—The obscure thermal rays also caused a strong excitatory reaction (Fig. 92). Attention is here drawn once more to the antagonistic reactions of temperature and radiation effects of heat.

It has been shown that the rays which cause the most intense excitations in *Mimosa* also induce the greatest retardation in the rate of growth. Thus ultra-violet is not only the most effective in causing excitation in *Mimosa* but also in retardation of growth. Next in order comes the blue rays: the yellow and red are practically ineffective in both the cases. Infra-red rays are, however, very effective in exciting the sensitive *Mimosa* and in retarding the rate of growth.

FIG. 92.—Effect of infra-red rays on the pulvinus of *Mimosa*.

DIVERSE MODES OF RESPONSE TO STIMULUS.

In *Mimosa* excitation is followed by the striking manifestation of the fall of the leaf. But in rigid trees contraction under excitation cannot find expression in movements. I have shown elsewhere that even in the absence of realised movement, the state of excitation can be detected by the induced electro motive change. I have shown that not only every plant but every organ of every plant is sensitive and reacts to stimulus by electric response of galvanometric negativity.*

There is an additional electric method by which the excitatory change may be recorded. I find that excitation induces a variation of the electrical resistance of a vegetable tissue.† Thus the same excitatory reaction finds

* BOSE—Friday Evening Discourse—Royal Institution of Great Britain. May 1901.

† This variation is sometimes positive, and at other times negative, according t the condition of the tissue.

diverse concomitant manifestations, in diminution of turgor, in movement, in variation of growth, and in electrical change. The correspondence in the different phases of response in pulvinated, ordinary, and growing organs may be stated as follows: Excitation induces diminution of turgor, contraction and fall of the leaf of *Mimosa;* it induces an incipient contraction or retardation of rate of growth in a growing organ; it gives rise in all plant organs to an electric response of galvanometric negativity and of changed resistance. All these excitatory manifestations will, for convenience, be designated as the *negative* response. There is a responsive reaction which is opposite to the excitatory change described above. In *Mimosa* the fall of leaf under excitation is due to a sudden diminution of turgor; the erection of the leaf is brought about by natural or artificial restoration of turgor. Rise of temperature induces an expansive reaction which is antagonistic to that induced by stimulus. Warmth also enhances the rate of growth and induces an electric change of galvanometric positivity.* The restoration of normal turgor or enhancement of turgor is associated with expansion, erection of the leaf of *Mimosa*, enhancement of rate of growth in a growing organ, electric response of galvanometric positivity, and contrasted change of electric resistance. All these will be distinguished as *positive* response.

There are thus several independent means of detecting the excitatory change or its opposite reaction in vegetable tissues. It will be seen that the employment of these different methods has greatly extended our power of investigation on the phenomenon of irritability of plants.

We have seen how essentially similar are the responsive reactions in pulvinated and in growing organs. It is therefore rational to seek for an explanation of a parti-

* BOSE—"Comparative Electro-physiology"—p. 75.

cular movement in a growing organ from ascertained facts relating to the corresponding movement in a pulvinated organ. The investigations on motile and growing organs that have been described fully establish the two important facts that, Direct stimulus induces contraction and Indirect stimulus induces the opposite expansive reaction. These facts will be found to offer full explanation of various tropic curvatures to be described in the subsequent series of Papers.

SUMMARY.

There is no breach of continuity in the increasing contractile reaction in a growing organ under increasing intensity of stimulus; the incipient contraction seen in retardation of rate of growth culminates in a marked shortening of the length of the organ.

Time relations of response, the latent period, the apex time, and the period of recovery are of similar order in pulvinated and in growing organs.

In condition of sub-tonicity the pulvinus of *Mimosa* responds to stimulus by an abnormal *positive* or erectile response. Under continued stimulation the abnormal positive is converted into normal *negative*. Growing organs in sub-tonic condition responds to stimulus by abnormal acceleration of rate of growth, which is converted into normal retardation under continuous stimulation.

Direct stimulus induces in *Mimosa* a *negative* response, with the fall of leaf. But Indirect stimulus induces the *positive* or erectile response. Similarly, Direct stimulus induces in a growing organ a *negative* variation, or retardation of rate of growth, and Indirect stimulus a *positive* variation or acceleration of rate of growth.

The electric response to Direct stimulus is by galvanometric *negativity*, that to Indirect stimulus by galvanometric *positivity*.

Under condition of sub-tonicity the autonomous activity of leaflet of *Desmodium gyrans* and of growing organs comes to a stop. The arrested activity in both is revived by the application of stimulus. Active pulsation in *Desmodium*, and active growth in growing organs are, however, retarded or arrested by stimulus.

The contractile effect of stimulus on pulsation of leaflets of *Desmodium gyrans* is seen by the reduction of the diastolic limit of its pulsations; to this corresponds the incipient contraction and retardation of rate of growth in a growing organ. The effect of warmth is antagonistic to that of stimulus. The expansive effect of rise of temperature is seen in *Desmodium* by the reduction of the systolic limit of its pulsation; in growth it is exhibited by an acceleration of the rate of growth.

All stimuli which induce an excitatory contraction and fall of the leaf of *Mimosa* also induce incipient contraction and retardation of rate of growth in a growing organ.

Excitatory effects of different rays of light on motile and growing organs are similarly discriminative. Ultra-violet light exerts the most intense reaction which reaches a minimum towards the less refrangible red end of the spectrum. Beyond this, the infra-red or thermal rays become suddenly effective in inducing excitatory movement and retardation of growth.